The Nature of (the) Universe

The Role of Negative Mass in the Infinite Space of a Timeless and Stable, Dual-Gravity Universe

By Herbert E. Taylor

The Nature of (the) Universe
by Herbert E. Taylor

Cover Design by Herbert E. Taylor (See Chapter 11)
Edited by Herbert E. Taylor
Graphics by Herbert E. Taylor

Published by Mountain Empire Publications
P. O. Box 480
Clifton Forge, VA24422

Printed and bound by Network Printers, Milwaukee, WI

Library of Congress Control Number 2006046900
ISBN 0-9707800-1-X

Printed in the United States of America

DEDICATION

I dedicate this book to my granddaughter, Patience Suzanne Smith, who just possibly might live long enough to see that ideas about the existence, characteristics, and role of negative mass in an infinite and stable universe have gained respectability in the scientific community.

Herbert E. Taylor

CONTENTS

ACKNOWLEDGEMENTS

The "seeds of thought" from which this book grew were planted over 50 years ago during conversations with a new-found friend, who like myself was a recent engineering graduate eager to start on a technically challenging career. But the start was delayed by the call to serve in the Army for two years. At that time, Jack Case and I were assigned to the same company at the U. S. Army's Engineer School, Ft. Belvoir, VA, and in short order established what has turned out to be a wonderful and enduring friendship. In our quest for intellectual stimulation to supplement the routine of army life, we met often after duty hours to go places, or do things, or simply to talk about wondrous ideas. Among other things, we visited the U. S. Naval Observatory in Washington, D. C. during one of its Open Houses and talked to a resident astronomer. Afterward we wondered about the newly-made claims about the "Big Bang" and concluded that "creation of the entire universe from nothing in an instantaneous explosion" was just unbelievable. We felt that there must be a better explanation about the nature of (the) universe. But the search for enlightenment on that matter got off to a slow start as Jack and I separated from the Army, pursued our own professional careers, raised our families, and kept in touch from opposite coasts of the country.

It was probably about 20 years ago that I began to notice more and more "glimpses" in the popular press of astronomer's and cosmologist's dissatisfaction with the "creation" and "endless universal expansion" aspects of the "Big Bang" scenario, and I began to develop some ideas of my own which I expressed in my letters to Jack. He encouraged me to write some more and over the years provided me with a lot of good resource material — so I wrote some more. After a while, it became apparent, as the thoughts and writing expanded, that they should be put into a book form. Once I got started toward creating a manuscript, I asked a few other friends to read and comment on early drafts of some chapters. They included Charles Fox, (the late) Edward J. McBride, (the late) Ammon Andes, and my daughter Leigh Taylor Smith. Their encouragement, along with Jack's continuing help, prompted me to keep going. I am especially indebted to Donna and Ed Sintz, Andre LaMontagne, my son Bryan C. Taylor, and Jack Case for reading and commenting on the "nearly complete" manuscript. I am also grateful for the technical help provided by Adam Richardson and Rudy Bondone to overcome computer problems involved in preparing the manuscript for publication.

Herb Taylor
30 Nov 2005

INTRODUCTION

I suppose that I should first explain that I chose my book title, *The Nature of (the) Universe*, before I ever became aware of the title of Fred Hoyle's book, *The Nature of the Universe*, which he wrote in 1950. And not coincidentally, that was about the time that I first took an interest, as a rank amateur, in the subject of cosmology. The thing that really set me off was the growing claim by some cosmologists that "the universe had been created from nothing," which I felt then, and continue to feel to this day, can't be right. In following years, I read a lot of interesting books and articles that I found in the "popular press," relating to astronomy and cosmology, but was too busy earning a living and raising a family to become a "scholar" on the subject. While my own ideas about the nature of a more believable infinite universe presented in this book have been developed over many years, it was only after my retirement from the "daily grind" that I seriously confronted the notion of writing a book. And it was only several years ago that I discovered Fred Hoyle's book, from a reference in some article. My initial reaction was disappointment in finding that the title of my book was "already taken." But after finding a copy of his book and reading it at this late date, I realized that he was undoubtedly the author of the ideas that I first picked up indirectly so many years ago challenging the "Big Bang Theory" (creation of the universe) — which he is credited with derisively naming. Over his working career, in conjunction with associates including Tommy Gold, Hermann and Christine Bondi, Geoffery and Margaret Burbidge, Jayant Narlikar, and Willy Fowler, ideas forming the basis for an infinite universe existing with "continuous creation" in a "steady state" were developed. I have learned about those ideas more directly from Sir Fred's last book written with Geoffery Burbidge and Narlikar, *A Different Approach to Cosmology*.

But their prolific output over many years seems to have been unable to overcome the hurdle of "evidence that the entire universe is expanding." And so, sadly, Sir Fred died in 2001 apparently without being able to advance his view of a steady state, infinite universe beyond one whose galaxies expanded continuously over some distant invisible threshold and disappeared into oblivion. But he describes how those lost galaxies can be replaced by continuous creation of "new matter" which evolves through stages from the primordial gas and dust to the elements that make up the more familiar solid matter of our world and galactic structure.

Rather than stealing from Sir Fred's title and ideas, I am unashamedly attempting to add another "brick" to the foundation that he and his colleagues have built, toward closing the gap between that "distant invisible threshold over which expanding galaxies disappear," and the primordial gas and dust from which they are recycled. I hope to describe how an infinite universe might work in some respect, not with the "disappearance" of galaxies, including those that might very well be expanding within one or more of its regions, but rather with the recycling of galactic material in a manner that includes a mechanism for balancing against the incessant attractive force of gravity that we know so well — a final act required to overcome the ridiculous idea that the entire universe must be expanding on a galactic scale always and everywhere. This is to affirm a belief that (the) universe must be infinite, and therefore, it must also be timeless and stable.

While my new and strange ideas might be easier to accept if I were a professional theorist in physics, astronomy or cosmology, I consider that the lack thereof frees me to add a fresh voice to the arena — being able to say obvious things without concern about losing a reputation, job or grant money. My background, and qualification to be heard, includes a unique mixture of academic training — Bachelor of Science and Master of Science in Mechanical Engineering, majoring in thermodynamics, heat transfer and heat power generation, along with teaching those subjects at the University of Kansas, and a career of practical engineering including design and development of equipment for supersonic aircraft and spacecraft, and industrial thermal process equipment — and a long term interest in the subject of cosmology. I also believe that my experience in "reverse engineering" (the art of figuring out how things work that you never designed or built), developed over a full career of confronting competitive products and systems, was valuable in my organization of thoughts expressed in this book.

1

GOOD QUESTIONS

Is there anyone who has ever stood in a spot of clear night air and looked up at the moon and stars without a sense of wonder? After all, almost every child, of English-speaking parents, has heard the familiar nursery rhyme/song "Twinkle, twinkle little star/ How I wonder what you are/ Up above the world so high…". But as wonderful as it is, most people go through life wondering very little, if anything, about the nature of the moon and stars — their relationship to our planet Earth, or the natural forces and processes which created, maintain and in due time will destroy them as those individually identifiable objects.

Fortunately, some children are more inquisitive about nature and begin to ask questions. For example, during a heavy thunderstorm, a child may ask what causes thunder. And a parent will likely answer that thunder is caused by "the clouds bumping into each other". Since most parents never think any more about the cause of thunder and because the answer satisfies the child's intellectual quest, the subject may never be discussed between them again; and the child's knowledge of the mechanics of thunder clapping may never advance beyond that point. I suspect that many parents really do know that there is more to it. They just don't understand it well enough to attempt explaining it to a child who has no concept of the generation of atmospheric static electrical charges and conductance through moist air, the transformation of electrical energy into heat and then heat into pressure producing sonic shock waves. But a child who maintains an interest in the subject and absorbs information over time from varying sources — perhaps even studies meteorology — will understand that there is more to it than most people want to know. Learning about the nature of the moon and stars — and of (the) *universe — too, can only follow the asking of intelligent questions.

How (the) universe works is one of several great philosophical questions. Some scientists seriously believe that it had a beginning (the "Big Bang" theory) and correspondingly that it will someday have an end in the form of a "big crunch" for an ultimately contracting universe (closed universe) or a "big chill" for a universe ever-increasingly expanding toward nothingness (open universe) or something in between for a universe forever expanding at a constant (steady) rate while somehow remaining intact (flat universe). The choice depends on how much mass and associated gravitational attraction each scientist thinks exists in (the) universe. Others believe that the existence of (the) universe transcends any measurement of time or space (Steady State theory — not to be confused with the steady expansion option of the "Big Bang" theory) and that time and space both extend to infinity, beyond anyone's ability to comprehend.

Those of us who believe that (the) universe exists in a steady state, claim foremost that there is no physical boundary to space surrounding us; that the farthest object which can be detected in outer space is not at the "edge of the universe" with nothing (no space) beyond. And (the) universe, in its infiniteness, did not come from anywhere and has nowhere to go. Finite objects ("things") come and go and are changed by the effect of external influences, e.g., growth by acquisition of mass or shrinkage by loss of mass, or property change by transfer of energy. But since there can be no exterior to an infinite space, there can be no change to the average condition throughout space. Though, of course, there is plenty of fuel for never-

---

* The use of parenthesis here and throughout the book signifies the meaning of an infinite universe as explained in Chapter 2

ending reactions creating local and temporary disturbances (from the norm) at any time and any point in space. In this regard, Stephen Hawking says "...we cannot give any particular reason for thinking [the universe] began one way rather than another. ... One could say: "'The boundary condition of the universe is that it has no boundary.' The universe would be completely self-contained and not effected by anything outside of itself. It would neither be created nor destroyed. It would just BE." (1)*

(The) universe, itself, must be as unchanging as any physical law which applies throughout, and any physical constant, such as the speed of light, or "alpha" (a parameter that combines the charge of the electron, the speed of light in empty space, and Planck's constant). Strangely, however, many people who are well educated and profess to scientific reasoning — the ones who normally form the "backbone" of support for logical thinking — apparently fail to embrace the concept of an infinite universe, either because they don't understand the meaning of "infinity" or reject the notion of its association with reality. E.g., "If (the) universe is infinite, it can't be real; but since (the) universe is real, it must not be infinite." In that lack of understanding, they are indeed caught in a "catch-22" whirlpool of thought. The thing to understand about the concept of infinity is simply that it extends beyond anyone's ability to comprehend. It is non-understandable, but it is still very real. The "shoreline problem," or Koch Snowflake, discussed in James Gleick's *Chaos* illustrates how the length of the shoreline around an island increases with the selection of smaller and smaller instruments to measure that distance. (2) The smaller the measuring instrument, the more attention must be paid to going around small bays and upstream little rivers and marshes feeding into the shore waters, before skipping over them at an arbitrarily set scale-factor. Imagine Paul Bunyan counting the number of paces around a large

---

* Numbers in parantheses relate to source in Reference section
   at end of chapter.

island and multiplying that number by his average distance per pace. Then imagine a flea, determined not to get its feet wet, doing the same thing. In the extreme, the shoreline length approaches infinity while the island and its enclosed area remain finite and very real.

Supporters of the "Big Bang" theory as an explanation of the "creation of the universe" have taken full command of the literary field of battle for ideas about the nature of (the) universe. Many reputations, jobs and grants for professionals in the field of Cosmology are apparently based on perpetuating the myth. Some, however, are beginning to look for an exit. In 1986, Dr. Andrei Linde, a cosmologist at Stanford University, advocated that the Big Bang was only one of many in a chain reaction of big bangs by which (the) universe endlessly reproduces and reinvents itself. "Any particular part of the universe may die, and probably will die," Dr. Linde said, "but the universe as a whole is immortal."(3) P. James E. Peebles recently wrote: "...You will notice that I have said nothing about an 'explosion' — the big bang theory describes how our universe is evolving, not how it began. We do not know what the universe was doing before it was expanding."(4)

Supporters of the Steady State theory have, for the most part, been reduced to simply claiming that the "Big Bang" didn't happen, and unfortunately, to the media, this view of (the) universe with such a non-event is non-news. But I have news. I believe that a big bang happened; it just wasn't quite the way that the "Big Bangers" claim. It didn't create (the) (whole) universe; it was just the latest event that re-created our region of (the) universe. And it was not a unique event; big bangs happen randomly in various regions of (the) universe from time to time and are a part of the natural mechanism that works to maintain timeless stability throughout (the) universe with an energy redistribution and gravitational balancing act.

The question about (the) universe isn't, how or when did it get this way, nor what's going to happen to it in the end. It has always been this way and will remain so forever. The question is, How does it stay this way — the way we see it now?

An understanding of how (the) universe manages to stay this way must include a plausible explanation of how it might work to maintain an overall gravitational stability, in spite of the relatively weak, but persistent force of gravity which we currently understand only in terms of attraction between bodies. And it must include an explanation about why it might not "run out of gas" by using up all available sources of heat energy, in spite of the Second Law of Thermodynamics that says no dynamic system, involving continuous transfer of energy from one form to another, can run forever on its own power.

Any hypothesis about the nature of (the) universe — how it works, whether or how it began and whether or how it will end — involves starting with a point of belief on blind faith, because ideas about the creation or non-creation of (the) universe are not scientifically testable. The trouble with any experiment to prove that (the) universe is or is not boundless, is that it would have to run forever before you could find the answer. Unfortunately, some people feel they must have an answer before that. I am reminded of the story about some disgruntled souls who are blaming the Food & Drug Administration (FDA) for discouraging drug companies from developing an immortality pill that would enable us to live forever, by requiring that its effectiveness first be proven by an actual test.

Those who promote the "Big Bang" theory as the model for explaining the nature of a universe which they see as a born and growing, evolving and expanding thing, would have you believe that the "entire universe" was created once-upon-a-time in a flash from nothing — a colossal leap of faith for a colossal event. If you were able to make that leap without questioning how such a thing might have happened (or better yet, if you rejected it as totally absurd), you should have no trouble at all in making the leap of faith required for serious consideration of my proposal about the nature of (the) universe — a tiny leap about the unusual behavior of tiny particles.

My proposal is based on the creation of mutant subatomic particles which are believed to make up an antigravity material called negative mass — particles that are most likely being created all the time in the bowels of the greatest of celestial furnaces, the giant neutron stars or black holes, but die quickly like most harmful viruses in the human body when the body has a strong, healthy immune system. But when the "immune system" of a celestial body weakens from excessive growth, the virus-like mutant particles multiply and eventually help to "destroy" the body by participating in the triggering of a significant structural change — a massive explosion which primarily acts to limit the size to which a giant body may grow. (There has to be something that keeps black holes from growing forever.) Secondarily and beneficially, it acts to redistribute energy and matter back into the spatial void, and restore gravitational balance to the surrounding neighborhood.

In support of this idea, and to de-bunk the "Big Bang" hypothesis, I argue that there are more believable, more mundane causes for the observations of the redshift in light waves that we receive from distant galaxies than those based on a belief that the "whole universe" is expanding as promoted by Edwin Hubble in 1929, and for the mysterious cosmic background radiation detected in 1964 and described as the "afterglow of the Big Bang" — both of which are cited by the Big Bangers as the solid foundations of their hypothesis. I will further argue that (the) universe can indeed operate as a perpetual motion machine of the second class (one which can run forever on its own power so long as it is never required to deliver power externally), rather than being ultimately doomed to expire as one big cool ball, or expanded into nothingness.

I, of course, can no more suggest an experiment that would give scientific proof to my ideas, than can anyone prove that a universe can be created from nothing. It will take a lot of work for the scientific community to refine the hypothesis and advance it toward the status of theory, through use of mathematics, computer studies and interpretations of celestial observations, to establish widespread acceptability of these ideas — if they should deserve it. Personally, I think that anyone who has had experience establishing acceptability for the "Big Bang" hypothesis would have no trouble with this simpler task.

My thesis develops primarily from the belief that space and time are boundless — space extending to infinity in all directions and time extending to infinity in the past as well as the future. No amount of paperwork, theories, mathematical formulations, computer printouts, or scientific publications, is going to prove whether they are or are not. It is purely a matter of choice of which idea seems most logical. My next choice from the menu of possible beliefs is that matter and energy fields are distributed uniformly throughout space, on average, without regard to time. These are pivotal choices in that if they are believable, one cannot also logically believe that (the) universe was "created". It should be apparent that if (the) universe has been created from a point a lá the "Big Bang" hypothesis, it would take an infinitely long time for the "created" matter and energy to fill the infinite space. Although, I read an article recently, now lost in the shuffle of papers on my desk, in which the author compromisingly commented that if space is infinite, then (the) universe could still have been created an infinitely long time ago. But that is to paradoxically claim that one can comprehend an event which happened at a point in time beyond the range of comprehension. We can't have it both ways. So, if we are to firmly believe that time and space are infinite and that (the) universe can exist forever as we now see it, we must look for the clues that support that belief and share them with anyone else who may be searching for landmarks along that same seldom traveled path.

QUOTATIONS

1. All men by nature desire to know.   Aristotle

2. As we acquire more knowledge, things do not become more comprehensible but more mysterious.
                                Albert Sweitzer

3. The first thing a man of understanding must understand is that there is much that he will never understand.
                                Jack Miles (5)

4. All truths are easy to understand once they are discovered; the point is to discover them.
                                Galileo Galilei

5. It is morally as bad not to care whether a thing is true or not, so long as it makes you feel good, as it is not to care how you got your money as long as you have got it.
                    Edmund Way Teale, *Circle of Seasons*, 1950

6. Infidelity does not consist in believing or in disbelieving; it consists in professing to believe what one does not believe. It is impossible to calculate the moral mischief, if I may so express it, that mental lying has produced in a society. When man has so far corrupted and prostituted the chastity of his mind, as to subscribe his professional belief to things he does not believe, he has prepared himself for the commission of every other crime.
                        Tom Paine, *The Age Of Reason*

7. Man has made machines that can answer questions, provided the facts are previously stored in them, but he will never be able to make a machine that will ask questions. ...The ability to ask the right questions is more than half the battle of finding the right answers.
                        Tom Watson/Howard Eves (6)

8. There are three great philosophical questions: ...What is life? What is consciousness and thinking and memory and all that? And how does the universe work?
                        Edward Fredkin/Robert Wright (7)

9. It is difficult to imagine that a handful of residents of a small planet circling an insignificant star in a small galaxy have as their aim a complete understanding of the entire universe, a small speck of creation truly believing it is capable of comprehending the whole.

    Murray Gell-Mann

10. The universe is not only queerer than we imagine, but queerer than we can imagine.   J. B. S. Haldane

11. ...It is far better to grasp the Universe as it really is than to persist in delusion, however satisfying and reassuring.

    Carl Sagan (8)

12. ...[T]heories of cosmology...cannot be tested with a controlled experiment. There is no way, except in our imaginations and our computers, to rewind the tape and play it again. We are confronted with a world, a fait accompli, and must work backward, piecing together a story of how it came to be. ...We must stop at some point, dig in our heels, and simply declare that our representation is valid. You cannot have a theory of representation — you can only represent.

    George Johnson (9)

13. There is a coherent plan in the universe, though I don't know what it is a plan for.   Fred Hoyle

14. (Ed Fredkin) believes that atoms, electrons, and quarks consist ultimately of bits — binary units of information, ...and he believes that the behavior of those bits, and thus of the entire universe, is governed by a single programming rule.   Robert Wright

15. (An interview with Sir Roger Penrose)...
    Q. There was a debate...last fall, "The End of Science". Do you agree with those who argue that when it comes to physics and mathematics we know just about all that can be known?
    A. The argument is that accessible important ideas will run out, that either we will never understand certain

things or we have already understood them. This is an absurd position to take. People have been saying that for centuries.   Claudia Dreifus (10)

16. The number of rival theories seems to be greater in cosmology than in most sciences, with the exception of psychology. In cosmology there also seems to be a tendency to do more publishing and less critical examining of theories than in most branches of science.
                                    Reginald Kapp (11)

17. Most cosmologists would be delighted to abandon the standard model [the Big Bang theory] for something new to think about if only the alternatives looked reasonably promising.   P. J. E. Peebles/ Nicholas Wade (12)

18. Using the forces we know now, you can't make the universe we now know.
                        George Smoot/Warren E. Leary (13)

19. For every complex problem there is a solution that is simple, neat, and wrong.   H. L. Mencken

20. It is important not to confuse the universe as it is with the universe as we wish it would be. ...No amount of theorizing is apt to converge on a persuasive explanation of where the mathematical laws are written or what happened before the Big Bang. For all powers to observe and reason, the mind ultimately encounters chasms. Then the only choice is to retreat or take the great leap and choose what to believe.   George Johnson (14)

21. (Edgar Allan Poe, in his book _Eureka_, 1848)...argued for a method to knowledge which he called imagination; we now call it intuition.... Imagination, or genius, or intuition, lets the classification start so that the successive iterations, back and forth between empirical and rational, hone the product until it eventually conforms to nature. Only then is the dross of the classifier skimmed away and a true order in nature, if it exists, revealed.

Simply start, and like Poe, trust in the imagination.
Allan Sandage/Timothy Ferris (15)

22. The practical value of astronomy...is to know about the objects in the heavens and their arrangement, but it is not clear that this knowledge will lead to anything that will directly affect our lives. Perhaps the greatest benefit will come from an unexpected direction. ...The finding of something that cannot be explained may lead to a new principle of physics which will greatly influence our lives. ...It is rare that an idea outside generally accepted lines of thinking is supported for further development. A new discovery can blast scientific thought out of a rut and into new directions.   Philip A. Charles (16)

23. The assumption that the universe can be reduced to an original particle has already changed or rather, degenerated into a second assumption, the myth of the Unified Theory. Many physicists are now inclined to believe that even if we cannot find the smallest building block of the universe, we can find a mathematical formula that will explain the entire universe: a Theory of Everything. ...The belief that the universe is "written in the language of mathematics" is not only wrong, it is entirely outdated. ...My argument is not simply that it is not given to humans to explain everything, including the universe. When human beings recognize that they cannot create everything and cannot see everything and cannot define everything, such limitations do not impoverish but enrich the human mind. They mark the evolution of consciousness.   John Lukacs (17)

24. How long can cosmology stay respectable when its practitioners have no idea what kind of matter makes up at least ninety percent of the universe?
Nicholas Wade (12)

25. ...It has become necessary to add one elaboration after another to the original Big Bang theory — the existence

of a brief "inflationary epoch" right after the initial explosion; vast amounts of invisible, unexplained "dark matter", and now perhaps some mysterious something that is accelerating the cosmic expansion. ...Some scientists have taken to calling cosmology the craft of making theories of the universe — cosmetology. There always seems to be another blemish to cover up.

George Johnson (18)

26. [About the missing mass mystery], I have a feeling that we just need to admit to ourselves that the emperor has no clothes. John N. Bahcall/John Noble Wilford (19)

27. Now cosmologists realize that things aren't so straightforward. The universe may not be governed by the gravity of ordinary matter after all, ... Matter has little to say of its own fate. Instead the universe may be controlled by the so-called cosmological constant, a surreal form of energy that imparts a gravitational repulsion rather than the usual attraction. If there is a story to be seen in cosmic history, it is the march from the utter simplicity of the big bang to ever increasing complexity and diversity. The near-perfect uniformity of the primordial fireball, and the laws that have governed it, have steadily given way to a messy, but fertile heterogeneity: photons, subatomic particles, simple atoms, stars, complex atoms and molecules, galaxies... . Understanding how the intricacy is immanent in the fundamental laws of physics is one of the most perplexing philosophical puzzles in science.

George Musser (20)

28. Galaxies are the fundamental structures into which matter is organized in the universe, yet we don't have a clear understanding of how they formed or how they evolved into the varieties we see today.

Chuck Seidel/John Noble Wilford (21)

29. ...The Infrared Astronomical Satellite found that superclusters of thousands of galaxies, interrupted by voids some 200 million light-years across, are common in the visible universe. Scientists do not believe the force of cold dark matter alone could have worked fast enough to create structures so large. Even 20 billion years is not enough time for thousands of galaxies to have clumped together in the way the theory says. ...For the most part, though, nature follows simple rules. So while cold dark matter may exist, astronomers are beginning to search elsewhere to solve the mystery of how galaxies were born
Michael D. Lemonick (22)

30. We clearly do not know how to make large structure in the context of the Big Bang... . Someday, we may find that we haven't been putting the pieces together in the right way, and when we do, it will seem so obvious that we will wonder why we had not thought of it much sooner.
Margaret J. Geller/John Noble Wilford (23)

31. On a larger scale...very massive objects weighing as much as a million Suns, might be drifting around in the halo (of the Milky Way galaxy), but if that were true, Dr. (Christopher) Stubbs said, their gigantic tidal pull would have yanked apart the global clusters of stars seen, apparently undisturbed, in many parts of the galaxy. ...The hidden mass might consist of some undiscovered form of exotic matter. It might be he said, that "gravity works differently from what we think."
Malcolm W. Browne (24)

32. ...Some, as yet unknown, force operates in the extragalactic space to keep the galaxies apart.   Reginald Kapp (11)

33. Jaan Einasto...says they have found reason to believe that superclusters — giant globs of galaxies — arranged in a gargantuan "three-dimensional chessboard", (extend) throughout the heavens. ...If true, this would be stunning news. There is little reason to believe that the Big Bang...

scattered its debris with more care than any other blast. A universe so fastidiously and geometrically arrayed would require... "some hitherto unknown process that produces regular structure at large scales" — in short, new laws of physics. ...Gravity creates its own order: the roughly spherical planets and stars, the whirlwind galaxies. Could there still be another force that arranged the superclusters into an array like the atoms in a piece of quartz? ...There is a price for presuming that everything can fit on science's theoretical canvas: we must be ready for these inevitable surprises, and willing to tear it up and start over again. George Johnson (25)

34. As far as astrophysicists can tell less than 1 percent of the mass of the universe has revealed itself through visible light and other detectable radiations. The rest cannot be detected directly, and is only inferred to exist from the gravitational effects it appears to exert on the galaxies. Scientists have been seeking this mysterious dark matter for decades.   John Noble Wilford (26)

35. [The failure of the Hubble telescope to find "hidden mass"]...seems to shift more of the burden of the search for missing matter from observational astronomers to particle physicists, who seek to explain the universe in terms of concepts drawn from theory and their atom-smashing experiments in powerful accelerators. Their current speculation is that the bulk of the universe may consist of strange hypothetical particles that have never been detected. ..."Our results increase the mystery of the missing mass. ...They rule out a popular but conservative interpretation of dark matter." ...If the missing mass was ordinary matter, then scientists were forced to consider exotic alternatives that existed only in theory. ...Determining the nature and amount of dark matter is one of the fundamental issues of cosmology.

John Noble Wilford (27)

36. All of the visible stars and galaxies, scientists now believe, account for less than 1 percent of the mass of the universe. The rest, then, must be some kind of invisible dark matter that has so far eluded detection by the best instruments of astronomy and particle physics....The first intimation that there might be more to the universe than accounted for in previous theory came in the 1930's. Dr. Fritz Zwicky, an astrophysicist at the California Institute of Technology, observed some galaxies traveling at unexpectedly high speeds in a cluster of other galaxies. Since the observable mass in the cluster was not sufficient to provide the gravity needed to pull galaxies at such high speeds, he decided there must be a great halo of invisible mass surrounding the cluster. ...Indeed astrophysicists may have to think of a few more impossible things before their next breakfast, if they are to figure out what the universe is made of. ...One more big discovery could change this. "One of us is going to find dark matter", Dr. (Charles) Alcock said. "That will be the great coup of the 90's in cosmology."   John Noble Wilford (28)

37. Progress in science is a continual to and fro between theorists and experimentalists. While the theorists sit with pencil and paper scribbling models of the universe, it is up to the experimentalists in the laboratory to find a way of testing these theories. ...The trouble is, it has been difficult to verify these new theories, because their predicted repercussions could not be tested by any known technology.   Simon Singh (29)

38. Andrei Linde...hypothesizes many...universes, perhaps an infinite number of them. The material world originated not with "our" big bang but with the first big bang, which probably happened an infinitely long time ago. "The big bang remains a very interesting theory that we must study, but it is somewhere in the distant past." He called the origin of our bubble universe "not the big bang, but a pretty big bang."   Timothy Ferris (30)

REFERENCES

1. Hawking, Stephen W. (1988), *A Brief History of Time*, Bantam Books, New York, p136

2. Gleick, James (1987), *Chaos*, Penguin Books, New York, p 99

3. Overbye, Dennis (2001, May 22), Before the Big Bang, There Was ... What? New York Times, p F1

4. Peebles, P. James E. (2001, Jan), Making Sense of Modern Cosmology, Scientific American, p 54

5. Miles, Jack (1996), *God, A Biography*, Vintage Books, p 291

6. Eves, Howard M. (1988), *Return to Mathematical Circles*, PWS-Kent Publishing Co., p 63

7. Wright, Robert (1988, Apr), Did the Universe Just Happen?, The Atlantic, p 29

8. Sagan, Carl (1996), *The Demon-Haunted World*, Ballantine Books, New York, p 12

9. Johnson, George (1996), *Fire in the Mind*, Vintage Books, New York, pp 265 & 316

10. Dreifus, Claudia (1999, Jan 19), A Mathematician at Play in the Fields of Space-Time, New York Times, p F3

11. Kapp, Reginald (1960), *Towards a Unified Cosmology*, Scientific Book Guild, London, pp 17 & 161

12. Wade, Nicholas (1994, Nov 27), Star-Spangled Scandal, New York Times Magazine

13. Leary, Warren E. (1990, Jan 14), Spacecraft Sees Few Traces of the Tumultuous Creation, New York Times

14. Johnson, George (1998, Jun 30), Science and Religion: Bridging the Great Divide, New York Times, p F4

15. Ferris, Timothy (1995, May 15), Minds and Matter, The New Yorker, p 46

16. Charles, Philip A. et al (1995), *Exploring the X-ray Universe*, Cambridge University Press, Cambridge UK, p 22

17. Lulkas, John (1993, Jun 17), Atom Smasher is Super Nonsense, New York Times, p A25

18. Johnson, George (1998, Mar 8), Once Upon a Time, There Was a Big Bang Theory, New York Times, p 3wk

19. Wilford, John Noble (1994, Nov 29), Astronomy Crisis Deepens as the Hubble Telescope Sees No Missing Mass, New York Times, p C1

20. Musser, George (1999, Jan), Getting Complicated, ScientificAmerican, p 6

21. Wilford, John Noble (1996, Feb 27), When the Young Universe Gave Birth to Galaxies, New York Times, p C1

22. Lemonick, Michael D. (1991, Jan 14), Bang! A Big Theory May Be Shot, Time, p 63

23. Wilford, John Noble (1991, Jan 15), New Surveys of the Universe Confound Theorists, New York Times, p C1

24. Browne, Malcolm W.(1995, Apr 18) Dark Matter Search Hints at New Shape for Galaxy, New York Times,p C1

25. Johnson, George (1997, Jan 19), The Real Star Wars: Between, Order and Chaos, New York Times, p 4E

26. Wilford, John Noble (1997, Apr 29), Scientists Report Discovering "Missing Mass" Component, NewYork Times, C3

27. Wilford, John Noble (1994, Nov 16), Space Telescope Leaves Scientists Puzzled About Mass of the Universe, New York Times, p B8

28. Wilford, John Noble (1992, May 26), Physicists Step Up Exotic Search for the Universe's Missing Mass, New York Times, p C1

29. Singh, Simon (1998, Jun 16), The Proof is in the Neutrino, New York Times, p A31

30. Ferris, Timothy (1995, May 15), Minds and Matter, New Yorker, p 46

2

(THE) UNIVERSE

"When I use a word…it means
just what I choose it to mean —
neither more nor less." Lewis Carroll

The prevailing concept among cosmologists, as seen in countless publications over the past many years is of a universe which has the attributes of a "thing" — having definite (though uncertain) age, size, and content (mass). "Things", of course, can be described by their properties, like mass, size, shape, color, texture, etc, and other parameters, such as energy levels, like pressure, temperature, electrical charge, chemical composition, etc, and age or "life span" (creation to destruction) — each by reference to some standard of measure (i.e., within limits). But I do not view "the universe" as a "thing" because I believe that it has no limits.

What we call this spacetime that we occupy for the moment is deserving of more distinction. And I believe there is an etymological problem which acts to stifle a better understanding of its nature that results from the invariable manner of reference to "the universe".

When we write and talk in good grammatical form about nouns, of course it is appropriate and common to use the word *the* prefixed to the referenced noun. According to Webster's dictionary, *the* is a definite article "used as a function word to indicate that a following noun…is definite or has been previously specified by context…(as in put the cat out)". Its use implies the existence of some limitation in the following

30

object (the noun), whether the noun be concrete or abstract. Used in conjunction with a concrete noun (like a cement floor — i.e. a thing), the implications of boundaries are obvious, as in mass or size (e.g., the cat, the house, the planet, the star, the galaxy or any other individually identifiable object). In contrast, abstract nouns or proper-named objects may be prefixed by the article *the* or not, depending on the desired meaning. For example, we may say that "love springs eternal" — which implies an unbounded concept of love; or we might speak about "the love of mankind" — which discounts all other forms or ideas of love except that applying to mankind. In a similar manner, even for some concrete objects, we omit the article *the* to imply a quality of abstraction. When I worked as a teacher in a public school, I would tell my wife when I left the house that I was going to the school (to work — a rather specific task in just a few subjects). But at the same time, if we were to speak about it, we would say that the children were going to school (to learn many subjects — a more open process). And on Sundays we went to church (to worship), but occasionally at other times when I had a limited duty (e.g. a volunteer maintenance job) to perform, I went to the church. Similarly, doctors go to the hospital (to work) while patients go to hospital (at least in the United Kingdom) to be healed.

Strangely enough we may talk about the earth or refer to (the planet) Earth just as we might refer to Mars (or Venus, Mercury, etc), but never to the mars (or the venus, the mercury, etc). And we refer invariably to the moon or the sun, but never to Moon or Sun. We may talk about planning a trip around Mars, but we do not say that we have already been to Moon. We may also say that we are going to take a trip through space, but never through universe (— around the universe, maybe).

Thus, the use of the word *the* can have a powerful effect on the meaning of an expression by implying a limitation on the identity or scope of a thing as being only a part of a larger thing. After the dissolution of the USSR, the country now known simply as Ukraine got rid of the Russian version of its name, Ukrainskaya (The Ukraine). Volodymyr Zholdokov, Assistant to the Ambassador at the Permanent Mission of Ukraine to the United Nations, carefully noted, "We insist on dropping 'the' before 'Ukraine', for the article implies here (that it is) part of some bigger country, which is not the case." (1)

Insidiously, however, the word *the* can also be easily overlooked or hidden as an operator. Since the word *the* is the most commonly used word in the English language, we tend to ignore its existence or are positively put off by it. Technical writers of product instruction manuals are often guilty of omitting the term altogether (e.g. "Press button on lower control panel to turn on motor.") for the sake of brevity and focusing attention on an object or action.

The wording of the original tobacco warning, "The Surgeon General has determined ..." was a masterpiece of ineffectiveness to convey an important message — a godsend to the tobacco industry. If readers weren't immediately turned off by the leading *the*, they were turned off by reference to the Surgeon General, a government official having until then very little public recognition — someone that nobody knew or about whom nobody cared. With this linguistic conditioning, it is no wonder that "the universe" is thought of as a "thing" — a component of a larger entity. There seems to be complete, though wrong, consensus in the current literature that it had a beginning (the "Big Bang" theory) and that it will have some kind of ending due either to gravitational attraction (or ultimate loss of it) or to "using up" all available heat energy.

The "Big Bang" theory places meaningless and unnecessary limitations on the basic concepts of time and space — as if it were possible to have a point in time at which there is (or was) no longer any time or space. Many scientists say that time and space are continuums and that phrases like "a point in time" or "a point in space" (a singularity) are in themselves meaningless. A limitation in size (i.e. expanse or volume) of "the universe" is implied by writers subscribing to the "Big

Bang" theory (a lá, the universe was once no bigger than a "point" and is now 15 billion light-years to "the edge of the universe"). Limitation of mass is frequently discussed in the literature in terms of the latest estimates on the "total mass" in "the universe", recently estimated at 125 billion galaxies. All the while, that amount is acknowledged to be only somewhere between 1% and 10% of the mass needed to satisfy calculations about the supposed Hubble expansion of space.

Other free-wheeling author/speakers corrupt the meaning of "the universe" by describing "multiple universes" (an oxymoron) as in "a different kind of universe that existed before the 'Big Bang'", or "what's on the other side of a black hole?", or countless "bubble universes" that grow within a larger and otherwise empty "multiverse", or "parallel universes", or "asymmetric universes", or "sequential universes" within which Mario Livio calls the "global universe" in his book, *The Accelerating Universe*, ad nauseam.

That said, I must now make it clear that I believe that celestial space which surrounds us has temporally existed and will continue to exist forever, and extends endlessly in all directions. Those few writers who seem to share my belief use "the infinite universe" to describe that concept. But I will be so bold as to claim that this is not only a redundancy in terms (infinite universe) but also oxymoronic since *the* implies that the object (universe) is part of something larger. What we need is a better, and shorter, standard of reference.

One possibility in clarifying the more abstract, infinite nature of "the universe" would be to drop the *the* and simply refer to "universe". But given the entrenched usage of "the universe", that would seem awkward and likely never catch on. So instead, I propose adding parenthesis marks, as follows, when referring to (the) universe. The parentheses seem appropriate for this purpose on two accounts. First, parenthesis marks are used in ordinary literature for side comments and in technical documents for secondary reference dimensions and instruction notes, that may optionally be ignored.

Second, parentheses are used in accounting records to designate a system having dual and balancing values — credits and debits, addition and subtraction, positive and negative. The parentheses are also appropriate when viewed from this balancing perspective. For those who firmly believe that (the) universe is infinite, the parentheses will help to keep focused on the concept that it is neither part of something larger, nor just one among many. And for those who agree with my ideas described further on, the parentheses are doubly appropriate in reference to a belief that (the) universe has a dual balancing system of gravity which acts galaxy-to-galaxy, alternately in time and alternately in space, to forever maintain the rough overall positional stability of the three-dimensional "chessboard" pattern of galaxies throughout space as we see it now.

QUOTATIONS

1. ...Most physicists, like Dr. (Abhay) Ashtekar, believe that since there is just one universe, there should be just one fundamental way of describing it.   James Glanz (2)

2. The tendency has always been strong to believe that whatever received a name must be an entity or being, having an independent existence of its own. And if no real entity answering to the name could be found, men did not for that reason suppose that none existed, but imagined something particularly abstruse and mysterious.   John Stuart Mill

3. You know, words mean more than we mean to express when we use them.   Lewis Carroll

REFERENCES

1. Nelson, Derek (1997), *Off the Map, The Curious Histories of Place Names*, Kodansha International, p 154

2. Glanz, James (1999, Apr 20), Taste-Testing a Recipe for the Cosmos, New York Times, p F1

3

THE SPHERICAL MOON

In this chapter, I will point out an example of how gravity works relentlessly toward a goal in natural perfection — the shaping of newly "created" celestial bodies into (nearly) perfect spheres. We readily understand the working of gravitational attraction from our own personal experience. When we stumble, we fall down. And we can project that experience into an understanding of how mountain peaks, when they fall down from asteroid bombardment (or earthquakes) and fill up the valleys, the rocks don't get up again by themselves. Attractive gravity is the only type we really know, of course, and the only type to be experienced in our, or any possible, solar system and throughout most of our or any other galaxy — at least, until the next big bang. However, we should pause and consider that after a big bang, gravity in a dual form acting between galaxies, might work just as relentlessly toward another goal in natural perfection — the positioning of newly created (re-created) stars into galaxies and then the newly re-created galaxies into a roughly uniform spatial gridwork pattern, relative to their older neighbors. That idea will be discussed in further chapters.

The sight of the full moon shining in a clear night sky evokes many different thoughts to different types of people — poets, lovers, song writers, farmers, sailors, hypochondriacs, members of the occult, astrologers and astronomers. The one thing we all see is a (nearly) perfect circular (disc like) reflection of sunlight. And it is interesting to me to contemplate the process of study by early man when they first took notice of the similarity in appearance between the moon during its cyclic display and that of a ball when it is spotlighted with illumination from different angles. The same variation in crescent shapes of light from nothing (backlighting) to full circle (front lighting), surely indicated that the moon was also a (nearly) perfect sphere. I can further imagine that the strength of that understanding emboldened some sages to feel confident, again by rule of similarity, that the land they stood on and the seas they sailed were parts of the surface of another spherical body — later understood to be the planet Earth. In fact, in Egypt the diameter of the earth was calculated to a surprising degree of accuracy by measuring the length of a pole's shadow at a precise moment when it was known that the sun was directly overhead a well hole (with no shadow at its bottom) many miles away, some seventeen hundred years before Columbus ventured to find a western sea route to the Indies. Circumnavigation of the spherical body, Earth, was later documented by Magellan.

Beyond that, I don't think that anyone besides an occasional astronomer or geologist is impressed very much by the remarkable sphericity of these and other celestial bodies. Like gravity, it is just seen to be there and taken for granted. But, of course, the moon as well as the earth wasn't always a (nearly) perfect spherical body. There are different theories about the origin (formation) of our solar system, the earth, and the moon. Geophysicists estimate that the earth as we know it today was formed about 6 billion years ago from an enormous galactic explosion (a supernova in the Milky Way) that flung material far and wide. Some of it settled into orbit around what is now our sun.

The moon is believed to have been created more than 4 billion years ago from the debris of a cataclysmic collision between Earth and another large, Mars-sized, body. (1, 2, 3) Knowing what rocks look like when dynamited out of the side of a hill, I doubt that neither the earth nor the moon were anywhere near spherically shaped "in the beginning" (after their respective big bangs) when the core pieces, each in its own time, fell into a recognizable orbit. They probably looked like gigantic pieces of jagged rock.

How did they become such (nearly) perfect spheres? Simple — through the action of gravity. Unrelentingly, the gravitational field of a celestial body attracts space debris (asteroids/ meteoroids) which is probably plentiful following a big bang. In the case of the moon, its small size produces insufficient gravitational force to hold an atmosphere, so that asteroids or smaller meteoroids slam directly into the moon's surface (becoming meteorites) without the attenuation afforded by the earth's atmosphere which burns up smaller meteoroids (creating meteors). Any given asteroid/meteoroid striking the moon's surface is just as likely to hit a peak as a valley — a random shot. If the meteorite ends up in a valley it becomes just another rock tending toward filling up that valley. If a peak is hit, and especially if it is a gigantic hit, even more material will be dislodged and fall into the valley. And a few real "blockbusters" (larger comets and asteroids) have apparently generated so much heat upon impact that it melted the crust, creating a "magma ocean" — a really level surface! Tests performed on "moon rocks" brought back from Apollo mission landings confirm that as much as 400 kilometers, about 250 miles, of the moon's upper crust was once molten. (1)

Now to illustrate the persistent, and powerful effect of gravity with this case, let us suppose that a rock the original size and shape of the moon could be "whittled" into the (nearly) perfect sphere that it now is by random collision with, say, a million meteroids/asteroids/planetesimals. A million hits to the moon over a time span of 4 billion years would average out to one every 4 thousand years. It is indeed humbling to imagine that the force of gravity is powerful enough to transform the moon into its (nearly) perfect spherical shape, with only single hits occurring over the span of the whole history of civilized man. But the moon will never be a perfect sphere, say, with a surface like a billiard ball. The same gravitational force which acts relentlessly toward its ultimate goal of forming the body into a perfect sphere also keeps the meteoroids coming in to blast out craters. This process makes the "texture" of rocky hills and valleys, forever superimposed upon the spherical shape that our astronauts had to avoid in finding a level spot to land the Lunar Excursion Module.

Incidentally, while few people ever give it a second thought, the everlasting broken-rock surface is also the reason that the full moon looks more like a shining flat disc hanging in the night sky rather than a sphere. The reflected light from the full moon is of almost uniform intensity across the whole face, from one side to the other — very close to the normal reflection of light from a flat disc (fig. 1). Light illuminating the surface of a perfect sphere, like a chrome-plated, polished ball will present a maximum intensity from the direct reflection at its center and taper off quickly towards the perimeter where the light is reflected more and more in the sidewise direction (fig. 2). The sun's light falling onto the hill-and-valley, rock strewn surface of the moon is actually reflected in all different directions from any given area. On a clear night, a small crescent of new moon appears with almost the same brightness for that given visible area as during a full moon, when the sun's rays are coming from a different direction but still reflected toward the earth. Since the moon's surface is more or less uniformly broken, the amount of surface inclined to reflect light directly towards the earth is also uniform, when illuminated from any direction (fig. 3).

However, since the period for the moon's rotation about its own axis (a lunar day) is just equal to the period of its orbit around the earth (full moon-to-full moon), and its axis of rotation is perpendicular to its orbital plane, the same face of the moon is always pointed towards the earth. This phenomenon is not just a coincidence, but also a result of gravitational force acting between the earth and the moon, or more precisely, the gradient of gravitational force. But for me to attempt to explain that any further here would lead me too far astray from my objective.

Thus, we see how gravity works to give a basic and simple geometric shape to newly "created" celestial bodies — a process that we have no reason to doubt is being executed with relentless repetition throughout (the) universe. Why should we not believe that gravity works just as well, perhaps in a new way to be understood, to relentlessly position newly "created" bodies, and groups of bodies, into a basic, simple and repetitive pattern throughout (the) universe in a stabilizing manner which can go on forever?

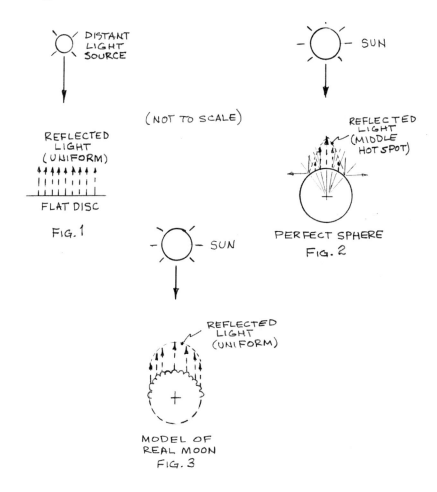

DISTANT
LIGHT
SOURCE

(NOT TO SCALE)

REFLECTED
LIGHT
(UNIFORM)

FLAT DISC

FIG. 1

SUN

REFLECTED
LIGHT
(MIDDLE
HOT SPOT)

PERFECT SPHERE
FIG. 2

SUN

REFLECTED
LIGHT
(UNIFORM)

MODEL OF
REAL MOON
FIG. 3

QUOTATIONS

1. Accepted wisdom in planetary science is that a rogue planet, the size of Mars (or 2-3X) struck the earth and dislodged some of its crust and mantle (as well as losing some of its own matter) — which remained in orbit around Earth and finally coalesced into the single large moon seen today.　　　　John Noble Wilford (2)

2. The new data (from Lunar Prospector spacecraft) reinforce the theory that the moon was created by a cataclysmic collision between Earth and another body more than four billion years ago.　　Mark Alpert (3)

3. That comets come in trillions verifies the planetesimal idea, for comets are but a species of planetesimal from uncrowded icy orbits. The densely pockmarked faces of the moon and the airless planets, moreover, record the waning barrage that closed the curtain on planet formation, a process that is now all but extinct (in our solar system.)　　　　Philip Morrison (4)

REFERENCES

1. Ward, Peter D. & Brownlee, Donald (2000), *Rare Earth*, Copernicus (Springer-Verlag), pp 50 & 54

2. Wilford, John Noble (7/29/97), Astronomers Recalculate "Whack" That Made the Moon, New York Times, p C6

3. Alpert, Mark (1999, Jun), The Little Spacecraft That Could, Scientific American, p 20

4. Morrison, Philip (1999, Jul), The Hidden Cosmic Ruckus, Scientific American, p 104

4

AN EXPANDING UNIVERSE?

Ideas that "the universe" was created in some specific manner and particular time were developed independently in many cultures. A high point of precision was reached in 1649 when the English Archbishop, James Ussher, proclaimed that according to the biblical record, the world and time began (by our current calendar), at 6 o'clock P.M., Saturday, Oct. 22, 4004 B.C. (1) But the "Big Bang" theory added a new wrinkle — that the whole universe was expanding, and that the expansion was proof of its creation (by extrapolation of the apparent expansion rate back to zero).

The "Big Bang" theory, about the "creation of the universe" starts with an unbelievably fierce event, and to many doubters of the theory, an event that is just patently unbelievable. Triggered by God knows what, an explosion is said to have occurred at no particular point in time or space, because time and space did not exist before the event according to the theory. The parameters of time and space are claimed to have been created about 15 billion years ago in our then-granted perception of time. Space is said to have blossomed at that instant from a point now referred to as the singularity point. It would not have been surrounded by space as we know it, because space is described in the theory as growing progressively outward as an expanding sphere, carrying with it all matter and energy now existing within "the universe," which was also created by the event. Since no supporter of the "Big Bang" theory seems to be taking an exception to belief in a basic law of physics regarding conservation of matter and

41

energy (i.e., the sum total of matter and energy within a closed system is a constant — "the universe" being an extreme case of a closed system) the matter and energy now distributed throughout "the universe," is described as having expanded to its current 15 billion light-year spherical radius size from an infinitely dense point "at the beginning of time".

The notion that space is expandable grew from Einstein's theory of relativity and Edwin Hubble's report of distant light sources whose radiation he saw as an indication of redshift. From that came the idea that the many light sources are not just speeding away from our observation point on Earth, but are all speeding away from each other and from any observation point within "the universe."

Hubble found a correlation between the degree of deepening redness in the farthest light sources and their distance from us which could still be determined by triangulation from opposite sides of the earth's orbit around the sun. Simply stated, he claimed that the farther a light source is away from us, the faster it is moving away. The ratio of recession velocity between two bodies and the distance between them is called the Hubble constant, H_0. The relationship has been widely accepted among astronomers and used by extrapolation to establish distances to light sources far beyond the range of determination by the triangulation method, although the whole process is fraught with controversy. The distance between Earth and a very distant light source is obtained by dividing the observed "recession velocity" (which is converted from a measured redshift and adjusted by comparison to a standard source of light from a dependable class of stars for corrected brightness from a known distance) by the Hubble constant.

Unfortunately, most of the present controversy among astronomers is simply about the correct value of the Hubble constant, currently estimated at 70 +/- 7 km/sec/10^6 parsec (2). (1 parsec = 3.26 light-years) and its implication for assignation "of the age of the universe," currently estimated at about 15 billion years. Additional controversy involves whether the apparent expansion of "the universe" is constant (flat universe), increasing (open universe), or decreasing (closed universe). Accordingly, the "Big Bang" theory offers the view that "the universe" was created from nothing at a singularity point, and has grown to its current size, which is not infinite, but may, or may not, grow indefinitely toward infinity.

The expansive growth of space is believed, according to this theory, to occur at the rate of the Hubble constant. However, if we can at least agree that "the universe" currently has a finite average density of mass/energy per cubic meter of space, and that the law of conservation of mass and energy is valid, then growth of such a universe to infinite size would result in its being filled with nothing, because density goes to zero as volume goes to infinity with constant amount of mass/energy.

Neil de Grasse Tyson describes an alternative view of universal expansion in his article, "Cosmic Horizons" (3). He also sees "the universe" as having been created at a singularity point via the "Big Bang" (which he places at 13 billion years ago) and expanding from there. But his "universe" which expands radially outward at the speed of light is contained within an apparently virtual infinite universe, which he calls a "meta-universe." The meta-universe must, in turn, contain a latent form of space/time and mass/energy which come into existence upon illumination by the expanding front of the light wave propagating from the singularity point of the "Big Bang," like the creation of an image on photographic film, or the hardening of dental cement by exposure to ultraviolet light. Thus, the singularity point is apparently not burdened with all of the accouterments of a full-blown universe, but mass and energy are picked up along the way.

Tyson says, "This cosmic sphere, with a radius of 13 billion light-years, is centered in space and time on whoever is making the observation. Moving at the speed of light, the expanding horizon continuously overtakes parts of the cosmos that were previously on its other side. In a billion years, when the universe is 14 billion years old, we will have seen another billion years of cosmic evolution at every distance between here and there, and we will have seen another spherical shell of cosmic material that has undergone the big bang and given birth to galaxies."

My personal view is more akin to, yet also slightly different from Tyson's. I believe that all the matter and energy is out there in an infinite space, but it was not created by radiation from a "Big Bang." I believe that big bangs occur periodically at random times and points throughout infinite space. And they are instrumental not in creation of matter and energy in the locality of the event, but in the recycling (re-creation) of matter and regeneration of thermal energy to restore a balance of gravitational forces and availability of high temperature thermal energy. More specifically, I can very well believe that a big bang occurred in our region of (the) universe thirteen, or fourteen, or fifteen billion years ago, and we may still be feeling the effects of expansion from that blast. I just don't believe that it was a unique occurrence, or that the associated local expansion defines (the) universe. An expansion in one region of (the) universe is surely reacted by contraction in adjoining regions. Thus, an expansion started in one region will not grow forever, but is simply one phase of a long term continuously oscillating process.

If we are to be convinced that "the universe was created" as the Big Bangers claim, I should think that it would be a great selling point for them to say, "Just look at all of those most distant stars that we have found, and are now seeing them as they were formed about a billion years after the "Big Bang." Isn't that enough proof?"

It does seem that just about every year for the past decade, we have learned about astronomer's sightings of new objects at record breaking distances.

A sampling of reports follows:

Dec '94 - Dr. Duccio Machetto claims to have found a "long-sought population of primeval galaxies", estimated to be 12 billion light-years away and having been created when the universe was only 2 billion years old, assuming a universe 14 billion years old. (4)

Mar '96- Dr. Allan Sandage concluded that seemingly ancient globular star clusters are 13 billion years old [created only about 2 billion years after the Big Bang], while the parent universe is at least 15 billion years old. (5)

Mar '98- Researchers at Keck Observatory in Hawaii found an unusual object that was 12.2 billion light-years away assuming the universe to be 13 billion years old. (6)

May '00- Astronomers from Sloan Digital Sky Survey located what may be the most distant object ever observed: a quasar about 13 billion light-years away, formed less than a billion years after the Big Bang, assuming the universe to be 14 billion years old. (7)

These are exciting reports, but taken as a group, there is a serious void in the information. To support the "Big Bang" theory, the next most valuable fact to know, after the great distance of each would be the same as the three most valuable pieces of information any realtor would advise you to consider if buying or selling a piece of real estate — location, location, and location.

I wonder if all of these most distant stars that are believed to have been formed within about a billion years after the "Big Bang" are indeed located near "the singularity point?" Simply put, if all of the most distant stars came from the same point within about a billion years, why hasn't anyone made news of the fact that they are all clustered within one or two billion light-years distance of each other? I suspect because they are not.

However, if someone were to tell me that all of the most distant stars have, in fact, been found within a celestial arc of about 8° (required for a cluster no more than 2 billion light-years across at an average distance of 14 billion light-years) I would have to argue that even that should not be taken as a solid point of support for the "Big Bang" theory. If we believe that big bangs, and re-creation of celestial bodies occur in a random pattern throughout (the) universe, then even a cluster of big bangs would not be highly unusual. If we were to flip a coin, say several hundred times, it would be very unusual if we did not encounter a string of five or six, or more, heads or tails in a row somewhere along the way. Indeed, I would expect that there would be a number of most-distant stars that were re-created in close proximity of some particular celestial point after the latest big bang in our region of (the) universe. But we should not fool ourselves into believing that all of the most-distant stars, whose estimated differences in ages may be less than the light- years of distance separating them, all came from the same big bang.

Charles Lineweaver and Tamara Davis describe in a recent article, Misconceptions About the Big Bang (8), a typical set of observations relating the time interval over which we are able to witness the illumination of a "standard candle" type of supernova and the distance to that supernova as indicated by redshift measurement, which is claimed to establish the "proof" of an expansion of space, not just movement of objects in space, but expansion of space itself, as follows:

- The duration of a nearby supernova is always about two weeks
- The duration of a supernova in a galaxy at a medium distance (redshift 0.5) lasts about three weeks.
- The duration of a supernova in a galaxy at a great distance (redshift 1.0) lasts about four weeks.

The understanding, of course, is that all of those supernova events are similar in brightness and duration at their place and time of origin, which is not a point of argument here. The question and argument is about the natural cause of the extension in the duration of our observations of the distant supernovae; i.e., how does two weeks of illumination at a distant origin transform into four weeks of illumination

46

received at planet Earth? Certainly, we perceive this as a significant difference, which basically, I believe is about the speed of light and not about an expansion of space to be explained later. It is a difference that must be viewed in its proper perspective. Here I am reminded of the following mental challenge.

> Assume that we put a metal belt around the Earth's equator (about 25,000 miles long) and draw it tight so that it is flush with the surface. Then we cut the belt and insert a piece one yard long into it, and allow the belt to separate freely from the surface, due to the slight expansion, uniformly all the way around. (Don't ask how it is held in that position this is just a mental exercise.) The question is, Will the belt be lifted enough for anyone to be worried about tripping over it?

Most people, not being mathematically inclined, do a quick comparison of the one yard addition to the length of the equatorial belt and conclude that since 36 inches is insignificant in comparison to 25,000 miles, the resultant rise of the belt would not be enough to trip anyone. But, of course, they would be wrong. Since $C = 2\pi R$, or $R = C/2\pi$ and $\Delta R = \Delta C/2\pi$, the lift of the belt would amount to $36/2 \times 3.14 = 5.72$ inches, which is truly insignificant in comparison to the Earth's radius of 4000 miles, but is quite significant in comparison to our height of 5-6 feet. Thus we see that something that makes a significant difference to us may result from a very insignificant difference in causative matters.

The point that I would like to make by description of an analogy is that what we see as a significant difference in durations for supernovae at varying distances from us, might result from a cause that is so insignificant that it is easy for most of us to be ignorant of and those who should know better to forget, and in doing so, ascribe the cause to something else.

My analogy, which may not be perfect in every respect, is done in two parts: First for the "expanding universe" as claimed by the Big Bangers, and second for a more reasonable alternative non-expanding universe. The analogy involves a load of sheep being hauled to market in a truck that experiences several accidents at different times and places, enabling the sheep to escape from the truck and run back home.

The set-up for the first half of the analogy is this:

- Load of sheep = Quantum of photons
- Accidental escape of sheep = Supernova event
- Motion (velocity) of truck at time of accidental escape = Expansion rate of universal space
- In all accidental events, it takes the same time for the sheep to escape from the truck, say, 10 min. (measured by the truck driver's watch).
- Home (destination of the loose sheep) = Earth
- Sheep are all programmed to run on the loose at the speed of light — 120 mph in a perfect vacuum, for this analogy.

1a) A truck loaded with sheep is driven away from home and one mile down the road, unbeknownst to the driver, the tailgate falls open and the sheep jump off the truck in waves over a 10 min. period. Each wave takes off running towards home as soon as their feet hit the ground. They all arrive back home over a period slightly in excess of 10 min., from the first wave of sheep to the last.

1b) The next day, the sheep are loaded up again and the truck is driven away. This time, after ten miles down the road, the tailgate comes open again, the sheep escape as before and run back home. Except this time, they all arrive back home over a period of 12.5 min., from the first wave to the last.

1c) The third day, the sheep are loaded up again and the truck is driven twenty miles before the tailgate falls open. The sheep escape again and run home, arriving over a period of 15 min., from the first wave to the last.

If we accept the premises of this much of the analogy — sheep always escaping over the same time period (10 min.) and running back home at only one speed (120 mph), then it is understandable that one might conclude that the difference in arrival duration for all the waves of sheep on different days was caused by the sheep jumping off the truck while it was still moving. Each following wave would have had to travel farther to get home than the one preceding it. And furthermore, for this set of data, the truck would not have been always moving at the same speed at the time of escape but going 30 mph at 10 miles away from home, and 60 mph at 20 miles away. Before proceeding with the second half of the analogy, we should review certain problems with the above conclusion as applied to the real world.

First of all, the idea that all of space within an infinite universe might be continually expanding is sheer nonsense (literally — lacking any sense or meaning). Second, to believe that light might travel all the way from a distant source, billions of light-years away, at only one precise speed, would be naïve. Light travels at 3.00×10^5 km/sec only in a perfect vacuum. It travels at a slower rate through any real medium — 2.24×10^5 km/sec through clear water (a reduction of 25%), and 2.00×10^5 km/sec through clear glass (a reduction of 33%). In the first part of the analogy, to accept the premise that sheep would always be able to run at the speed of light in a perfect vacuum, is to suppose that every last one of the sheep will always have a clean shot all the way home on a wide open road. But celestial space is not all a perfect vacuum. It is filled with "stuff" (referred to in the next chapter about redshift), which averages 10^7 atoms/cu. meter in interplanetary space within a solar system, 5×10^5 atoms/cu meter in interstellar space within a galaxy, and 1 atom/cu meter in intergalactic space. In this regard, celestial space should be characterized more accurately as a very, very thin but lumpy "soup." Here, we should stop and think about how much reduction in speed of light would be required to delay a photon of light by two weeks during a travel of one billion light years (redshift 1.0). $2/52 \times 10^9$ = $.04 \times 10^{-9}$ = 0.000000004% reduction.

To model the nature of celestial space more realistically for the second half of the analogy, we will add some obstacles on the road home which any individual sheep might encounter by chance, in which case its speed would be temporarily reduced causing it to fall behind one, or two, or three waves, or maybe arrive home just bringing up the rear. For this analogy, we will say that the obstacles — mud-filled potholes — are randomly located, but average one per mile. Any number of sheep may, by chance, encounter any one pothole, and any one sheep may encounter, in turn, any number of potholes located in its path.

- Pothole = any transparent media in celestial space; i.e., gas or dust clouds.

2a) The truck is loaded with sheep and is driven only a mile before it has a flat tire, necessitating a stop and fix, during which the tailgate falls open, the sheep escape in waves over a 10 min. period and they all run home as fast as they can (the speed of light for whatever media they encounter). As it happens, there is a mud-filled pothole which a few of the sheep encounter, that slows them down while they muddle through it and causes them to arrive home behind the sheep that missed the obstacle. In this case, since there was just one short delay and few sheep involved, their arrival period back home, from the first wave to the last of the stragglers, was just slightly in excess of 10 min.

2b) The next day, the sheep are loaded up again and the

truck is driven 10 miles before being stopped by another flat tire. The sheep escape again and run for home. But this time there are more potholes in their path, raising the probability that a pothole would be in the way of at least one sheep before it gets home, and the probability that the delay may be extended by any one sheep encountering more than one — maybe all — of the potholes. All of the sheep arrive back home, from the first wave (which has been slightly reduced by virtue of some of the sheep falling behind) to the last of the stragglers, over a 12.5 min. period.

2c) On third day, the sheep are loaded up again and the truck is driven 20 miles before being stopped by another flat tire. The sheep escape again and run home as fast as they can. But this time, there are even more potholes in their longer path home, which raises the probability that even more sheep would encounter and be delayed by potholes, and that the overall delay would be extended even more due to the higher percentage of sheep encountering multiple potholes. All of the sheep arrive back home, from the first wave (which is even more reduced due to more sheep falling behind) to the last of the stragglers, over a 15 min. period.

I believe the latter description of events gives a better understanding of the variable duration of supernovae observations, without having to invoke a meaningless, expanding universe. Willem de Sitter, professor of astronomy at Leiden in Holland, derived a solution to Einstein's 1916 General Theory of Relativity that proposed a naturally static universe, neither contracting nor expanding, with no need for an extra term of a Cosmological Constant. But for de Sitter's universe to remain stable, Einstein pointed out, it would have to be empty not even a single star. De Sitter argued that his model was viable as long as the distributed density of mass throughout the universe was negligibly low; i.e., amount of matter small compared to volume. (1) I would add, also, that from the mathematical point of view in an equation balancing forces, a universe filled with positive and negative mass bodies (galaxies) whose gravitational attraction to each other balance out to zero, would amount to the same as no mass at all — emptiness.

QUOTATIONS

1. The persistence of the disagreement [about the Hubble constant and missing mass] does worry scientists that somehow they are overlooking the most subtle mistakes in their observations or calculations, or else that their understanding of the cosmos is limited in some unrecognized but fundamental aspect.

    John Noble Wilford (2)

2. If in fact, the universe is closed, ultimately leading to a big crunch, our estimate of the age of the universe would refer only to the time since the most recent big bang. One can imagine an endless series of big bangs and crunches in an oscillating universe of possibly infinite age. ...How much time do we have before the big crunch begins? That is very difficult to say, because we don't even have proof that the expansion is decelerating at a fast enough rate to lead to an eventual collapse, and, in fact, the present best estimate says it is not.

    Robert Ehrlick (9, p 237)

3. Astronomers say... that the big bang was not an expansion of matter into pre-existing space, in which case there would be a real center and edge, but that space itself was created during the Big Bang.

    Robert Ehrlick (9, p 92)

4. It makes no more sense to say that the universe began at a certain moment in time than it does to say that it began at a certain point in space.

    George Johnson (10, p173)

5. The continuing uncertainty about the value of Hubble's constant, H_0, makes it difficult to state cosmological distances and time scales precisely. It is, however, a reflection of maturity on the part of cosmologists today that the uncertain scaling factor h_0 is found in most cosmological literature.   Jayant V. Narlikar (11)

6. In the late 1980s it was discovered that a large number of galaxies are not moving away from us in the uniform manner that big bang theory predicts. To account for this, theorists predicted...the existence of a conglomeration of mass, several hundred million light-years in size, that they called the Great Attractor. Other astronomers found evidence of a giant chain of galaxies: the Great Wall.   George Johnson (10, p76)

7. Either accepted stellar ages are too high or some aspects of the Big Bang theory are incorrect. It may be that something about the fundamental nature of the universe remains undiscovered, or possibly, all three are true. An older universe would help to explain how all the large structures, such as clusters of galaxies extending like great walls across more than a billion light years could have developed... .   John Wilford Noble (12)

8. ...[I]f nothing else, the crisis in cosmology demonstrates that there is much more to understanding cosmic history than immediately revealed in the Big Bang theory, however well established it may have become... . Dr. John Barrow...wrote: "One must understand that the term "Big Bang Model" has come to mean nothing more than a picture of an expanding universe... . The job of cosmologists is to pin down the expansion history of the universe — to determine how the galaxies formed: why they cluster as they do: ...and to explain the shape of the universe and the balance of matter and radiation within it." Dr. [Joseph] Silk, in his new book [*A Short History of the Universe*], conceded that the creation story told through the Big Bang theory might in a thousand years

be regarded as a 20th century myth, like the creation stories of antiquity. He hastened to add, "I am an optimist who finds our current paradigm so compelling that I can only imagine it will eventually be subsumed into a greater theory without losing its essential features."   John Noble Wilford (13)

9. Among those unhappy with the Big Bang theory were Fred Hoyle, Hermann Bondi and Thomas Gold. Over a period of many years culminating in 1948, these physicists formulated a rival theory that not only did away with the cosmic expansion, but in fact, accounted for many of the universe's observed features better than the Big Bang did. In their Steady State model, matter was continuously being created in the empty space between the stars and galaxies. While the notion of continuous creation struck many Big Bangers as absurd...the Steady Stater's arguments are quite formidable. The detection of cosmic microwave background radiation in 1964-65...was [taken to be] clinching physical proof in favor of the Big Bang. Although the Steady State model retained some true believers [and]...is not dead.   Helge Kragh (14)

10. There are parts of the universe, perhaps infinite in extent that we cannot see, because their light has not yet had time to reach the earth. ... There are always regions of the universe...that are inflating. They surround pockets of space where inflation has ended and a stable universe has unfolded. ...J. Richard Gott III and Li-Xin Li...recently proposed that the universe is trapped in a cyclic state, rather like a time traveler who goes back in time and becomes her own mother.   Martin A. Bucher et al (15)

11. The farthest one might conceivably see is out to the point where the recession velocity becomes equal to the speed of light, 3×10^5 km/sec. The light travel time from this point is 2×10^{10} yr [@ H_o =50 km/sec/10^6 pc]. This

is the reciprocal of the Hubble constant and is called Hubble time. Since the speed of light cannot be exceeded, information cannot be received beyond this horizon. ...The time elapsed since the Big Bang is the 'age' of the universe, and this age is dependent on the properties of space. If, for example, space is 'flat' the time since the Big Bang is 2/3 of Hubble time or 15 billion years ago. The age is uncertain because the average density of matter, which determines the 'curvature' of space is unknown. Present estimates of the age of the universe fall between 10 and 20 billion years.

Philip A. Charles (16)

12. ...[D]oubly special relativity predicts that the speed of light could depend on its color and energy. Such an affect might be spotted by observing gargantuan stellar explosions known as gamma ray bursts, says Lee Smolin... . The gamma rays take billions of years to reach Earth, Smolin says, giving the faster ones time to pull ahead of the slower ones.   Lee Smolin/ Adrian Cho (17)

REFERENCES

1. Gorst, Martin (2001), _Measuring Eternity_, Broadway Books, New York, pp 11, 226

2. Wilford, John Noble (1994, Dec 27), Astronomers Debate Conflicting Answers for the Age of the Universe, New York Times, p C1

3. Tyson, Neil de Grasse (1999, Mar), Cosmic Horizons, Natural History, p 90

4. Wilford, John Noble (1994, Dec 7), Hubble Captures Photos of Universe in Its Infancy, New York Times, p B9

5. Browne, Malcolm W. (1996, Mar 5), Age of Universe Is Now Settled, Astronomers Say, New York Times, pC1

6. Leary, Warren E. (1998, Mar 13), Light From Tiny Galaxy Reveals Most Distant Object Ever Seen From Earth, New York Times, p A16

7. Greenwald, Jeff (2000,' May 14), Brightness Visible, New York Times Magazine, p 24

8. Lineweaver, Charles H. and Davis, Tamara M. (2005, Mar), Misconceptions About the Big Bang, Scientific American, p 41

9. Ehrlick, Robert (1994), *The Cosmological Milkshake*, Rutgers University Press, New Brunswick NJ, pp 92\ 23

10. Johnson, George (1996), *Fire in the Mind*, Vintage Books, New York, pp 173 & 76

11. Narlikar, Jayant V. (1983), *Introduction to Cosmology*, Jones & Bartlett Publishers Inc. Boston, p 395

12. Wilford, John Noble (1994, Oct 27), Theory of Universe's Age Poses New Cosmic Puzzle, New York Times, A1

13. Wilford, John Noble (1994, Nov 1), Big Bang's Defenders Weigh Fudge Factor, A Blunder of Einstein's, As Fix for New Crisis, New York Times, p C1

14. Kragh, Helge (1996), *Cosmology and Controversy*, University Press

15. Bucher, Martin A. & Spergel, David N. (1999, Jan), Inflation in a Low Density Universe, Scientific American, p 63

16. Charles, Philip A. & Seward, Frederick D. (1995), *Exploring the X-ray Universe*, Cambridge University Press, Cambridge (U.K.), p 326

17. Cho, Adrian (2005, Feb 11), Special Relativity Reconsidered, Science, p 867

5

REDSHIFT

The dubious ideas of an expanding universe and associated "beginning of time" (embodied in the "Big Bang" theory of the universe) apparently grew from a prediction based on Einstein's theory of relativity (1916) and observations by Edwin Hubble and Milton Humason at the Mount Wilson Observatory in 1929 which were interpreted as support for that theory. The two astronomers looking through their new and more powerful telescope began to notice that the light from more distant hitherto unseen stars appeared in a more reddish color than they were used to seeing in the Milky Way and closer galaxies. They concluded that the change in degree of redness was not just a consequence of those stars being more distant but was symptomatic of the phenomenon called redshift. Redshift is a change in radiation wavelength from one value measured at its source to a higher value measured by a distant observer who is moving radially away from the source. (I.e., the light is observed as shifted in the visible light spectrum toward the longer wavelength, red, end. Conversely, light seen by an observer moving toward its source is shifted toward the blue end — blueshift.) This phenomenon is similar to the more familiar acoustical (sonic) Doppler effect that causes a train's warning horn to first sound above pitch as it speeds toward a crossing, and then suddenly drop below pitch as the locomotive passes you at the crossing and speeds away.

Technically, the cosmological redshift is not a normal Doppler shift. The Doppler redshift and the cosmological redshift are governed by two distinct formulas. The first comes from Einstein's special relativity, and the second comes from general relativity, which can take into account an expansion of space. The two formulas are nearly the same for nearby galaxies but diverge for distant galaxies. (1) For illustration, the redshift effect is shown in fig.1 as it is applied to the detection and analysis of angular velocity of a star seen to be rotating about an axis perpendicular to the line of sight.

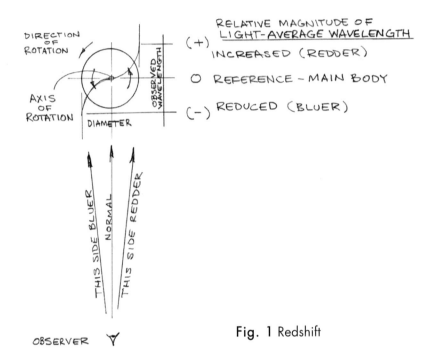

Fig. 1 Redshift

However, if the star itself is moving radially away from you, the entire light disc will most certainly be redshifted. But just looking at a lone star that might or might not be moving toward or away from you will not tell you whether the light has been "shifted." Determination of "shifted" light requires comparison of the received light's spectral "fingerprint" to that measured at its source, which is impossible to do in a direct manner for distant stars. However an indirect manner was found.

Among the billions of stars and galaxies now to be seen, there are many "light sources" within our own galaxy and beyond, that are recognized by their special pattern of light pulsation to be similar in make-up and at a point in their life span when the intensity of radiation (light output at the source) is reasonably the same for all of them (like a specific model of flashlight with brand new batteries) — and have been designated "standard candles." Their spectral distribution ("fingerprint") and intensity of emitted light is established by the chemical elements involved in the light-generating reaction taking place within the stars. The distance to the closest of these standard candles, within or near our own galaxy, can be determined by triangulation (measurement of the direction to the star from opposite sides of Earth's orbit around the sun). Farther out in space, where the difference in direction to a star from any place in Earth's orbit is too small for our instruments to measure, the distance to it is determined by measuring the intensity of the received light and comparing it to the intensity of those similar stars at a closer, known distance (by the Inverse Square Law — light intensity is inversely proportional to the square of the distance from the source).

Hubble's and Humason's most famous contribution to Cosmology was to find and measure the radiation from a number of those special light sources. What they measured basically, was that, within those distances which could be determined by triangulation, the intensity of light varied inversely with the square of the distance (as expected), but the spectral "fingerprint" wavelength increased (redshifted) pretty much in direct relation to the distance (surprise!).

Now it is a fact that the only proven mechanism for the shifting of wavelength in a beam of light transmitted directly from a source to a sensor, is by radial movement of one relative to the other. So of course, Hubble and Humason concluded that since those stars, scattered more or less randomly around space, were apparently moving away from us at a velocity increasing with the distance away from us, as indicated in the table below (2), the "whole universe" must be expanding — forming a "solid foundation" for the "Big Bang" hypothesis.

Cluster nebula in constellation	Distance, Light-years	Redshift, recession velocity, km/sec
Virgo	0.8×10^9	12×10^3
Ursa Major	1.0×10^9	15×10^3
Corona Borealis	1.4×10^9	22×10^3
Bootes	2.5×10^9	39×10^3
Hydra	4.0×10^9	61×10^3

As the light gathering power of our telescopes increased for viewing of objects deeper and deeper into space, the redness alone was taken as the sure sign of relative velocity induced redshift and related distance from us — now approaching 15 billion light-years — with an associated recession velocity approaching the speed of light. At this point, I would like to focus on some of the "bricks" (details/assumptions) built into the Big Banger's "solid foundation."

First of all, Einstein's equations of general relativity can be solved in a variety of ways — each one providing a model of how (the) universe might be. De Sitter's solution to those equations indicated, among other things, that the wavelength of light emitted from the far distant stars would appear to lengthen (redshift) as a result of the particular properties of time and space in his model (not because the stars were moving away from each other). The redshift would be in relation to the distance through which the light traveled — more distance, more redshift. (3)

So, it is a valid question to ask if redshift may also be caused by something other than the relative velocity effect. If so, it is not only not a sure indicator of relative velocity of far distant objects, but also may not even be a sure indicator of distance to far objects.

Hoyle, Burbidge and Narlikar describe in their book, *A Different Approach to Cosmology*, the initial finding of one of many quasi-stellar objects (QSOs) that appeared on a film at the edge of a galaxy. The redshift measurement of the QSO was much higher than that for the galaxy, indicating that the QSO was much farther away than the galaxy. The first impression was that there was no connection between the two. If anything, it was thought to be a chance of optical alignment, like those pictures of a tourist (in the foreground) pressing outstretched arms against the Tower of Pisa (in the background), to keep it from falling over. But after that, many more similar pairings of high redshift QSOs and low redshift galaxies were noticed — far too many to be the result of coincidental optical alignment. So after closer examination of other similarities in those pairings, it became an inescapable conclusion that the pairs of QSOs and galaxies were connected, and therefore had to be at the same distance from us (and presumably not receding at different velocities). Of course this conclusion has been attacked and "...one argument made against the reality of the associations by a leading observer was that if these results were correct, we had no explanation of the nature of the redshift! In other words, if no known theory is able to explain the observations, it is the observations that must be in error!" They further conclude "...that many QSOs have large redshift components that are not of cosmological origin." (4)

Furthermore, we observe that all galaxies within our view are not moving away from each other. There are many reports in the literature of galaxies merging, colliding, or in some cases just passing through each other. And I wonder, if the expansion of galaxies (movement away from each other) may be claimed by the Big Bangers as an indication that "the universe" was "created" 15 billion years ago, then what is indicated by the merging of galaxies — that that region of "the universe" has not yet been created? This is a tricky question as the Milky Way galaxy itself is reported to have consumed a dwarf galaxy sometime within the last billion years, and is now on a collision course with the Great Andromeda galaxy! (5, 6, 7, 8, 9)

Instead of a universe filled (somehow) with galaxies everywhere moving away from each other, as claimed by the Big Bangers, I visualize (the) real universe as being populated by bodies and groups of bodies moving in more or less chaotic directions under the influence of momentum and ever-changing (over long periods of time) attraction or repulsion from gravity or electromagnetic force fields. (Picture a roomful of angry bees and the relative motion of each one from any given single bee that you might call "Earth.") I believe that within the visible region of (the) universe, there should be according to some random distribution, very few bodies moving directly away from us and very few bodies moving directly toward us. Most of them should be moving in intermediate, random and non-intersecting directions. Relative motion, of course, can be resolved (divided) into vector components of radial (co-linear) motion and tangential or "astrometric" (perpendicular and non-intersecting) motion. There is, in fact, no fixed reference frame in space against which tangential motion can be measured. And if it could, what would the Big Bangers claim as its significance?

If, in fact, there is validity to the redshift causation <u>primarily, or only</u>, because of the relative velocity effect, and if it is logical to expect that there are just as many distant objects in (the) universe that are likely to be moving toward the earth as moving away from it, I would also expect there to be just as many reports from astronomers about observations of blueshifts. But I have found no reports of blueshifts. And that is evidence enough to me that the observed redness of the more distant objects may be caused by something other, or more significant, than the relative velocity effect.

At this point, I am happy to concede that most of the galaxies in the proximate region of our own Milky Way probably are moving away from us, still recoiling from the most recent big bang that happened here 15 billion years ago, and the redshift of light from those (that are receding) is primarily due to the relative velocity effect. (Allan Sandage has calculated that it will take 82 billion years, based on the current expansion rate and mass in "the universe," from the last big bang until it is ripe for the next one.) But I think that it is a credible hypothesis that the perceived redshift of the most distant light sources may be the result of the "weakening" of light by penetration farther and farther through space filled with "cosmic dust."

I am fully aware that the "tired light" explanation for the extra-redness of light from the most distant objects has been put forth for a long time, only to be rejected by physicists because the relative velocity effect is the only mechanism that has been "laboratory tested and certified" as capable of producing a true redshift of light. But I think that it deserves some more consideration. Indeed, the very meaning of "comparing" the properties of light received here and now to what it was at its source billions of years ago and billions of light-years away, is stretched by the reality that it can <u>never</u> be directly tested and measured in our Earth-bound (or even, Solar System-bound) laboratories.

I wonder if instead, the more distant, redder stars are "just naturally" redder due to the fading of light across so great a distance — an effect that is too weak to measure and "certify" within our own time and space, but becomes apparent after a few billion light-years — and in the extreme is more significant than the relative velocity effect. Like the cooling of a blue-hot billet of iron taken out of the furnace, to white color, then orange, then red and finally no longer incandescent, all things normally deteriorate with time. Can it be rightfully claimed that light, which has attributes of mass, can exist for billions of years and travel through billions of light-years distance without giving up some of its energy to the environment through which it travels? Intergalactic space is not a "perfect vacuum." So-called "empty" space is actually filled not only with detectable elementary particles and others that pop in and out of existence too quickly to be measured directly, but with other hard-to-detect things — escaped stars, dwarf (brown methane) galaxies, and superheated gas. And some authors add "dark matter" and an ocean of "vacuum energy" or "dark energy" to that list.

The following is an estimate of the "average density" of celestial space:

- Interplanetary space (e.g. solar system) 1×10^7 atoms/cu. meter

- Interstellar space (within a galaxy) 5×10^5 atoms/cu. meter

- Intergalactic space 1 atom/cu. meter

I remain skeptical that the observed extra-redness of the most distant light sources is "proof" of an "expanding and created universe." I find it more reasonable to believe that such observations may be the best (and only) proof that we'll ever have for another cause of "redshift" — if not a consequence of de Sitter's time and space, then extended travel through "cosmic-dust" filled space. And I continue to seek reinforcement for explanations as to how (the) universe must be working in a timeless and stable manner.

QUOTATIONS

1. Throughout the century, scientists have had to rely on maddeningly oblique methods, laden with assumptions, for measuring the size of the universe. They've had to guess, from purely theoretical considerations how bright a star or galaxy really is. Then from its apparent brightness, dimmed by the journey of the light through space, they judge its distance. ...More guesswork comes in when astronomers try to judge how much of the dimming of...light comes not from distance but from intervening cosmic dust.   George Johnson (6)

2. ...Astrophysicists said that the apparent absence of TeV radiation from most of the [gamma ray] bursts would not come as a complete surprise — even if they emitted it — because the universe is not entirely transparent to such energetic radiation.   James Glanz (10)

3. According to Paul Marmet and Grote Reber [a co-initiator of radio astronomy], quantum mechanics indicates that a photon gives up a tiny amount of energy as it collides with an electron, but its trajectory does not change. As the photon travels, its energy declines, shifting its frequency to the red. [However] Marmet has calculated...that a density of 10×10^3 atoms/cu. meter is required to achieve the observed redshifts. ...Such a high density would create a closed universe with a radius of only a few hundred million light-years.

Eric Lerner (11, p 428)

4. The Andromeda galaxy and we in the Milky Way galaxy are heading for each other at 300,000 miles per hour. ...Astronomers had...believed such events to be rare, although there had been clues to the contrary. ...In 1983 a satellite that looks at the sky at infrared wavelengths found that most star formation was occurring where galaxies were colliding. (Stars just beginning their lives are often obscured at optical wavelengths. They glow brightly in the infrared.) ...At least some galaxies have populations of globular clusters. In a galaxy known as NGC 7252, there are globular clusters made up of stars 15 billion years old, 500 million years old and only 10 million to 20 million years old. The last two are presumably the work of collisions. ...The dwarf galaxy known as Sagittarius...may have passed through our galaxy ten times. John P. Wiley, Jr. (8)

5. There are, in fact, a number of possible explanations of the Hubble relation other than the Big Bang. In a small corner of the infinite universe that we observe, ...[an] explosion of this epoch some ten or twenty billion years ago, sent plasma from which the galaxies then condensed flying outward — in the Hubble expansion. But this was in no way a Big Bang that created matter, space and time. It was just a big bang, an explosion in one part of the universe. Eric Lerner (11, p 52)

65

6. When we use redshift as a gauge of further distance, we are assuming the truth of the big bang. Without the theoretical framework, the individual observations would be meaningless.   George Johnson (12)

7. ...[Lauer and Postman] found that in addition to moving with the general expansion of the universe, the Milky Way and the nearby universe appear to be drifting...with an average velocity of 425 miles per second in the direction of the constellation Virgo. This motion [is] in a completely different direction from that inferred from other observations... . ...If expanses of the universe as big as a billion light-years in diameter are still drifting with respect to the larger universe, the two astronomers said, then the universe has structure of matter on much larger scales than predicted by current theory. ...Dr. Vera Rubin said "The universe is more complex than we thought." ...Visible matter is thought to make up less than 1 percent of the universe's mass, nearly all the rest being in the form of the mysterious dark matter, which must be largely responsible for the shape and motions of galaxies and clusters of galaxies.

John Noble Wilford (13)

8. One of the most significant unresolved issues in observational cosmology is whether the redshifts of quasars arise from the expansion of the universe. An objective assessment of the data still leaves a reasonable measure of doubt about the validity of the cosmological hypothesis.   Jayant V. Narlikar (2)

9. Light [has been detected] from a tiny misshapen galaxy...[which] is the most distant object ever seen from Earth (estimated 12.2 billion light-years). ...Dr. James R. Graham said...the new galaxy probably merged with other infant galaxies and matter to form a bigger, more mature one. ...It appears as patches of light and not as individual stars, indicating that gravity was pulling

together clumps of scattered matter to form the galaxy.

Warren E. Leary (7)

10. Dr. Roger Windhorst [says]...that galaxies grew by starting out as clumps of stars... [which] merged and collided, consolidating into the large galactic structures seen today. John Noble Wilford (14)

11. On April 6 [2000], astronomers from the Sloan Digital Sky Survey...located what may be the most distant object ever observed: a quasar (13-14 billion light-years away). ...To us, the most distant [quasars] look like intense points of infrared radiation. This is because space is scattered with hydrogen atoms (about two per cubic meter) that absorb blue light, and if you filter the blue from visible white light, red is what's left. On its multibillion-1ight-year journey to earth, quasar light loses so much blue that only infrared remains.

Jeff Greenwald (15)

12. When light moves through a medium, its wavelength suffers some distortions, leading to effects such as bending in water and the separation of different wavelengths, or colors. These effects also occur for light and particles moving through the discrete space described by a spin network. ... For any radiation we can observe, the effects of the granular structure of space are very small ... [but] these effects accumulate when light travels a long distance. Lee Smolin (16)

REFERENCES

1. Lineweaver, Charles H. and Davis, Tamara M. (2005, Mar), Misconceptions About the Big Bang, Scientific American,42

2. Narlikar, Jayant V. (1983), *Introduction to Cosmology*, Jones and Bartlett Publishers, Inc., Boston, pp23 & 395

3. Gorst, Martin (2001), *Measuring Eternity*, Broadway Books, New York, p 226

4. Hoyle, Fred, Burbidge, Geoffery & Narlikar, Jayant V. (2000), *A Different Approach to Cosmology*, Cambridge University Press, pp 122,134,327

5. Musser, George (1999, May), Here Comes the Suns, Scientific American, p 20

6. Johnson, George (1999, Jun 6), Building a Cosmic Tape Measure, New York Times, p 6wk

7. Leary, Warren E. (1998, Mar 13), Light From Tiny Galaxy Reveals Most Distant Object Ever Seen From Earth, New York Times, p A16

8. Wiley, John P., Jr. (1998, Apr), A Space Invader is Here, Smithsonian, p 20

9. Tyson, Neil de Grasse (1999, Jun), Between Galaxies, Natural History, p 34

10. Glanz, James (1999, Oct 26), Celestial Gamma Rays Open Windows on the Heavens, New York Times, p F2 10.

11. Lerner, Eric (1991), *The Big Bang Never Happened*, Times Books Div. of Random House, New York, pp 52 & 428

12. Johnson, George (1996), *Fire in the Mind*, Vintage Books, New York, p 68

13. Wilford, John Noble (1994, Mar 21), Milky Way Gets a Tug Way Out There, New York Times, p A13

14. Wilford, John Noble (1996, Sep 15), Looking Back 11 Billion Light-Years, A Glimpse at Galaxy Birth, New York Times, p A18

15. Greenwald, Jeff (2000, May 14), Brightness Visible, New York Times Magazine, p 24

16. Smolin, Lee (2004, Jan), Atoms of Space and Time, Scientific American, p 74

6

THE MYSTERIOUS BACKGROUND RADIATION IN SPACE

As early as the 1930s and 40s, with the advent of ever-increasingly more sensitive microwave antennae, a slight background noise was noticed, but was initially thought by astronomers and physicists to be natural and of no special importance. It may rightly be said that A. McKellar was the first astronomer, in 1941, to identify the excitative radiation as black body, and the temperature required for it was 2.3 K. But this finding became another "war casualty" and was essentially lost in observatory files for many years. (1)

The detection of microwave radiation became more interesting to astronomers. And the Big Bangers, especially, developed a line of thought in the 50s and 60s as follows:

a) There should be some evidence of the "Big Bang" in the form of background black body "relic" radiation throughout the universe.

b) A "Big Bang" mighty enough to "create the whole universe" would be the only way that such radiation could have been created.

In 1964, Arno Penzias and Robert Wilson picked up the mysterious low-level hissing noise through the microwave antenna at Bell Laboratories in New Jersey, and found it to be essentially uniform, regardless of the antenna's direction and persisting even after roosting pigeons and their droppings were cleaned out of the antenna. Radioastronomers at Princeton University were consulted and apparently for lack of any better explanation they concluded that the electronic noise heard by Penzias and Wilson must be coming from outer space — fossil radiation from the "Big Bang." That explanation was quickly accepted by many cosmologists as another piece of convincing evidence for the "Big Bang" theory.

However, I propose an alternate explanation — one which surely cannot be any wilder than the presumption of an instantaneous "creation of the universe." The nature of light has been explained and proven experimentally both on the basis of wave theory (Thomas Young in 1801) and photon theory (Albert Einstein in 1905). While the two theories seem to be at odds in some experimental cases, there is little doubt in the success of each theory in many others. And, of course, it is difficult in those cases where they are at odds to be more firmly convinced of one theory than the other, or to identify the "dividing line" where they should be separated.

I would like to discuss a puzzle about light, though it must be understood that while I will refer to the visible range of the electromagnetic spectrum, it might well apply to all those phenomenon in the non-visible range as well (infrared, ultraviolet, x-ray, radio, etc).

The Danish astronomer, Romer, was the first to measure the speed of light in 1675. He did that by observing variance in the time between eclipses of a Jupiter moon, viewed from the earth during Earth's nearest and farthest positions from Jupiter as they both orbited around the sun, and determined the speed to be about 3×10^8 meters/sec. More recent and accurate measurements establish the speed of light propagation in a vacuum to be 2.997925×10^8 meters/sec. (2) It is the same for all colors of light and is independent of the intensity and the relative velocity between the source and the sensor. The speed of light is denoted in mathematical terms by "c" and is included in many physical theories since it is believed to be the top limit of speed at which anything, from spacecraft to bytes of information, can travel. We observe that when a light is "turned on" there is a delay before the event can be detected elsewhere; the farther away, the longer the delay, and the delay in time correlates predictably to the speed of light. Light generally radiates from its source in all directions and its intensity diminishes according to the Inverse Square Law — intensity is inversely proportional to the square of the distance from the source.

Light, which is regarded both as a form of energy (wave theory) and a form of matter (photon theory), can be visualized to progress as a single ray propagating through space after the source is "turned on," as illustrated in fig. 1 — purely a simplification for discussion purposes.

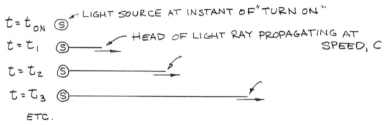

Fig 1 Light propagation after "turn on"

Light can travel from its source to great distances (theoretically to infinity) but, of course, its intensity is greatly reduced over long distances. Astronomers currently have telescopes with enough light gathering power to observe stars which they claim are 13 billion light-years away. (1 light-year is the spatial distance through which light propagates at speed, c, in 1 year — about 6 trillion miles.) So it is generally understood that when we look at the stars — even those close enough in our own galaxy to be seen with the naked eye — the light that we see is not indicative of the star's existence now, but of its existence many years, perhaps billions of years ago.

I occasionally encounter a comment in an article about distant stars which raises the question as to whether or not some of those stars might actually be extinct by now, and what we are seeing is merely an "afterglow" of history. And that is what puzzles me. Can light continue to exist in space independent of connection to its source, i.e., after the source has been "turned off"? Certainly light energy can be transmitted and stored in a sense, such as by excitation of fluorescent materials, or by a "solar cell" photoelectric generator. But that's not quite the same as saying that light itself can continue to exist without a source — like the residual stream of water from a garden hose which continues by its own momentum after the nozzle is shut off, and is ultimately absorbed into the environment.

If one were to argue that light cannot exist without a source, then it would also have to be argued that an entire ray (sphere) of light must de-propagate (collapse) at the instant that the source is switched off. That argument would imply that it's okay for light to take its time in propagating out into space, but it would have to de-propagate in no time at all. And that instantaneous de-propagation would constitute the transmission of a signal (information) from the source to the leading head of the ray(s) faster than the speed of light. But, if Einstein's theory of relativity is correct, this is not possible. "O.K.," says Steven Wright, "so what's the speed of dark?"

Instead, we must conjecture that de-propagation of light also progresses from the point of the now-missing source out into space at the speed of light, creating a distinct quantum of light no longer connected to its source, as illustrated in fig. 2.

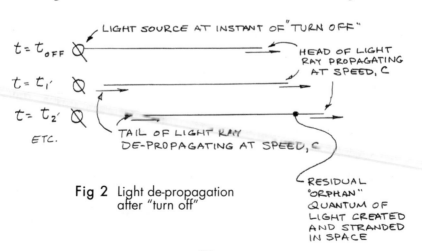

Fig 2 Light de-propagation after "turn off"

Now, if every time a light source is switched on and then off again, an orphan quantum of light is left adrift in space to dissipate, what happens to it? Theoretically, the orphan quantum of light might continue its outward travel to infinity with accompanying reduction of intensity to zero, and by that means "disappear" from (the) universe. But would it really disappear like that? Perhaps there is more to it. Any good physicist or mathematician knows that infinity and zero are merely concepts which are reachable only in a theoretical sense. In a practical sense to say that something real (e.g., light and distance) is zero or infinity is simply an admission that it is beyond our ability to measure more precisely.

My suspicion is that the mysterious background radiation derives from orphan light wandering endlessly throughout (the) universe. And if there is any credence to that, it might next be appropriate to consider whether the measured level of radiation (2.7 K) should be expected to increase as more lights are turned on and off in (the) universe, over time. Personally, I do not expect the level of background radiation to increase with time as a cumulative effect. If lights have been switching on and off forever, then it is not reasonable to expect that after another eon or two, the level will be any higher than it is now. I believe that the background radiation in space is at a stabilized level between the continuous creation of orphan light throughout (the) universe and another continuous mechanism for reprocessing it from radiant energy back into thermal energy (heat) — perhaps simply by collision of photons with "cosmic dust."

Hoyle, Burbidge and Narlikar give an elegant, academic presentation (way over my head) of an explanation for the cosmic microwave background radiation, in their book, *A Different Approach to Cosmology*. Basically, they contend that the microwave background comes from thermalization of radiation (starlight) from galaxies. The existence of the microwave radiation does not necessarily have anything to do with a "Big Bang," nor is there any significance of the value of the temperature (2.7 K) as far as the "Big Bang" is concerned. The temperature of 2.7 K is derivable from an analysis of a reasonable mechanism of transmission of starlight across billions of light-years, which is degraded into black-body radiation through absorption by common "cosmic dust" particles — long thread-like carbon whiskers — and re-emission at the temperature of the whiskers. Then a second-step absorption by metallic thread-like particles dominated by iron places the re-emission in exactly the region of the spectrum as that measured for the cosmic microwave background radiation. (1)

QUOTATIONS

1. The wave nature of light has been demonstrated experimentally as long ago as 1801 by Thomas Young using the famous "two slit" apparatus. In 1905 the quantum hypothesis was bolstered by Einstein, who successfully explained the photoelectric effect in which light energy is observed to displace electrons from the surfaces of metals. To account for the particular way this happens, Einstein was compelled to regard the beam of light as a hail of discrete particles later called photons.

   P. C. W. Davies et al (3)

2. In 1948, George Gamow, Ralph Alpher and Robert Herman predicted that if the universe had begun with a big bang, space would be permeated with its afterglow, in the form of measurable background radiation.

   George Johnson (4)

3. Modern cosmology depends on the prevailing inter-pretation of the microwave background — namely, that the microwave photons emerged when the universe was less than 1,000th its present size and traveled in a straight line for some 15 billion years. The strongest evidence for this is the nearly perfect thermal blackbody spectrum of the microwaves. This spectrum is easily explained if the radiation is primordial: it is thermal because the

photons were in thermal equilibrium at that early time. ...An alternative explanation — such as the "tired light" theory... would require a grand conspiracy to obtain a thermal spectrum.   Martin A. Bucher (5)

4. Is the microwave background, the so-called strongest evidence for the big bang, really of primordial origin and universal in character? Both its large- and small-scale isotropies are a problem under canonical interpretation. The former has the horizon problem and the latter the difficulty that no imprint of galaxy formation is observed on the radiation background. This latter point also brings out the fact that as yet no satisfactory under-standing exists of how galaxies formed in the standard models.   J. V. Narlikar (6)

5. The rapid variation of quasars implies that we have in some of these objects a switch which can turn the energy output of 10 billion suns on and off within an hour or two.   Freeman Dyson (7)

REFERENCES

1. Hoyle, Fred, Burbidge, Geoffrey & Narlikar, Jayant V. (2000), *A Different Approach to Cosmology*, Cambridge University Press, pp 79, 313, 314

2. Taylor, Edwin F. & Wheeler, John Archibald (1963), *Spacetime Physics*, W. H. Freeman and Co., San Francisco

3. Davies, P. C. W. & Brown, J. R. (1986), *The Ghost in the Atom*, Cambridge University Press, p 2

4. Johnson, George (1996), *Fire in the Mind*, Vintage Books, New York, p 69

5. (Letters to the Editor) (1999, May), Scientific American, p 8

6. Narlikar, J. V. (1983), *Introduction to Cosmology*, Jones & Bartlett Publishers Inc., Boston, p 394

7. Dyson, Freeman (1988), *Infinite in All Directions*, Harper & Row Publishers, New York, p 167

7

CREATION OF (THE) UNIVERSE?

The claim that "the universe" was created is a very old one. The earliest oral and then written records of human thinking consistently include accounts (legends) of the beginning of the narrator/author's "universe." The accounts naturally vary from one culture to another — e.g., Middle European, Far Eastern, Eskimo, American Indian, African Bushman, Australian Aboriginal. The desire to believe to know something about the origin of one's environment is a common human trait, though the substance of that belief is inevitably constrained by the narrator/author's comprehension and knowledge of the scope of the culture's environment. The "universe" of a South Seas Island culture, for example, might consist of only one island, the surrounding water and the observable sky above. The cosmology of that culture would thus involve the creation of that island, the water and space overhead. Similarly, an early account of an Eskimo culture would involve the beginning of ice and snow, bare earth and water (in season) and space overhead. The origin narratives become increasingly complex as modern cultures adjust and expand them to incorporate widening comprehension of their "universe," such as the awakening to the fact that their "island" is but a small part of a planet, and their planet is but a part of a solar system, and their solar system is merely an insignificant part of a galaxy, which in turn is virtually insignificant in relation to many other observable galaxies.

There is now a fantastic amount of material written by many people about the beginning of their "universe," which contains controversial hypotheses about how "the universe" began. These include the Creationist Theory about how a white-bearded, elderly, Caucasian male (a lá Michelangelo), named God, created it all (except himself, apparently, and whatever resources he had to work with) in seven days pretty much as presented in the Bible, and the Scientific Theory which describes it all as growing out of a "Big Bang," followed by a long process of inflation and evolution. Those theories are contrasted by the Steady State Theory which claims that a beginning of cosmic time has no meaning, and the challenge is not to explain how it all began, but to explain how it continues to exist.

While there may be many truths contained in these legendary accounts about the beginning of a particular "universe" (islands, planets, solar systems, galaxies, or beyond), they fail to provide me with a convincing explanation of the nature of (the) universe — much less that it had a beginning. The various arguments about how and when "the universe" began strike me as similar to arguments in the early Christian church about whether a particular feather relic was given by the Angel Gabriel or was taken from him, without, of course, carefully analyzing and determining if the feather might not have even come from an angel in the first place.

Concerning the nature of (the) universe, Edward Fredkin states "...There is a paradox that crops up whenever people think about how the universe came to be. On the one hand, it must have had a beginning. After all, things usually do. Besides the cosmological evidence suggests a beginning: the big bang. Yet science insists that it is impossible for something to come from nothing; the laws of physics forbid the amount of energy and mass in the universe to change. So how could there have been a time when there was no universe, and thus no mass or energy?" (1)

As a supporter of the Steady State theory, I consider the phrase "creation of the universe" to be an oxymoron. I can understand that the idea of living in a universe that was "never created" may give some people a very uncomfortable feeling — as though they were being shortchanged — deprived of a story about how things came to be. As a present-day product of an evolutionary process of intellectual and scientific learning, we are entitled to believe that while we may never agree about an "origin" of (the) universe, that at least our superior mentality will eventually enable an understanding of a process for its continuing existence.

It's not easy to want to believe that there can never be a logical explanation for a possible beginning of (the) universe. I recall reading an article about some explorers who recently succeeded in visiting and communicating with a remote jungle tribe, which had little or no previous contact with anyone from the outside modern world. The natives' needs and material possessions were very basic and simple, and for every man-made object that they had, if they had not made it themselves, they knew who did. And when the natives asked the explorers who made their fancy clothing and equipment, they could not understand that the explorers did not personally know who made their things. In addition to knowing where everything came from, the natives probably had a pretty good story about who "created their universe," though it was not in the article. And from this, I am reminded of what little progress has been made in the thinking of many "modern" individuals who feel that something is terribly wrong if they do not have a good story about who made "their universe". The concept of a non-created universe certainly runs counter to everything else that we know about (observe) in nature. There is an observable, or at least an imaginable life span for everything from subatomic particles whose existence approaches immeasurably short intervals to stellar galaxies billions of years old.

Before proceeding with my discussion of the non-creation of (the) universe, I would like to state explicitly that I have no dispute whatsoever with the body of cosmological evidence which indicates that there was a big bang involved in the "creation of our universe" — as long as we understand this quoted expression to mean the re-creation (i.e., recycling) of our region of (the) universe. My dispute is thus with the semantics or connotation of this terminology. I don't believe that the Big Bangers are talking about the same kind of creation or the same kind of universe that the Steady Staters are talking about.

Insofar as cosmological evidence increasingly confirms that a big bang was involved in the formation of our region of (the) universe, I say "hurrah." Personally, I believe that big bangs in various parts of (the) universe, from time to time, are a healthy and integral part of the continuing (steady state) process of nature. But to the idea that such evidence supports a conclusion that (the) universe was created from a single big bang, I say "nonsense." It is my hope to live long enough to witness a report that our depth of observation into outer space has surpassed the big bang region. (A hurricane spinning across an ocean looks different from a satellite above it, than from a ship inside of it.)

To me it certainly makes sense to inquire and study about the history of formation of our region of (the) universe. In dealing with such questions as, Where did "it" (i.e., mass/energy contained in our region of space) come from?, the consideration that it came from a transformation of mass/energy migrating into our region from adjacent regions, seems most plausible — like the swelling of the tide at a local seashore point. And with the acceptance of such an interpretation that our region of (the) universe had such a "beginning," doesn't it also follow that our region will sometime in the far unknown future experience an "end" (i.e., migration of mass/energy out of our region)? And if such a cycle can happen once, couldn't it happen again, and couldn't it have happened before, not only in our region but in other regions of (the) universe as well, because it is the pattern of nature to operate in cycles of action and reaction?

To pose the same question, Where did it come from? about (the) universe, merely leads us into "blind alleys." If we understand (the) universe as being of infinite extent, then there can be no adjacent region from which mass/energy can migrate. If the contention is that the mass in our (infinite) universe was formed by transfer of energy residing earlier therein, would it not be fair to ask then where the energy came from, and suggest that it probably resulted from an even earlier decay of mass distributed throughout (the) universe? At any rate, the premise that all matter and energy might have initially come from nothing, seems even less plausible than the possibility that (the) universe might have gone through one or more cycles of containing nothing except pure energy and pure mass at sequential points in time. The transformation of mass into energy is at least an understandable process (to some); the creation of something from nothing is more typically associated with magicians — like pulling a rabbit out of a hat.

Now that I have discussed some concepts of the "beginning" of (the) universe, I would like to discuss some related concepts of "beginnings" or "creations" of more common things.

For us, to argue that anything might have a beginning or an ending (life or death, creation or destruction), requires first of all that we know what we're talking about. I.e., there must be some way to reach agreement that we are talking about the same thing by identifying the subject. Beginnings and endings then are recognized as events relating to the identified subject. The life span of any physical thing is bounded first by its being recognized as an entity and then by the loss of that recognition — no longer matching a memory-stored image. The "creation" of the entity results from a process of transferring energy and agglomeration of mass which existed previously in the environment, and its "destruction" results from diffusion back into the environment. Physical laws hold that mass and energy may be transformed (according to Einstein's equation, $E=Mc^2$), but the total amount of combined mass and energy in any closed system remains constant. Identity then, is not the product of creation from and destruction to nothing.

An example of the way "beginning" and "ending" terms are commonly used may be seen as they relate to the discussion of a river. On viewing a physical map of any country, you can see that all major rivers "end" at the point where their waters empty into an ocean, sea or large bay. Following back upstream, a particular river may appear to "begin" from the outflow of a lake, or at the confluence of two minor streams, or simply trail off in a manner of uncertainty, leaving it a matter of local contention as to which brook or spring might be the real "beginning" of the river. Obviously, at either terminus, the water keeps flowing; the difference in identity between river and non-river is a matter of collective recognition and punctuation of complex systems. Similarly, any living thing is first "created" by a process of reproduction and at some point marking its "beginning" (recognition as a separate living entity), it is said to have been born. At a later point in time marking its "ending" (loss of recognition as a living thing), it is called "dead" and the body returns to "dust."

In the subatomic realm, life spans of particles are so short and elusive that the mere act of observing and attempting to identify and measure them has an effect on the results (the well-known Heisenberg uncertainty principle). Edward Fredkin argues that particles are (an identifiable) configuration of bits, not a certain kind of bits — that atoms, electrons, and quarks are nothing more than transitory patterns of information. They are not objects, as identified, in themselves and nothing more — they are things which behave like "electrons" at one time, or like something else at another time. (1)

Certainly at the other extreme, the estimated life spans of planets, solar systems and galaxies grow increasingly and incomprehensibly long.

For my part, I am satisfied that anything and everything that we can identify as an entity by placement of boundary conditions, can and will have a beginning and an end (i.e., be recognized as an entity at one time and be no longer recognizable as that same thing at a later time), however short or long its life span might be.

(The) universe, however, as a concept of space extending infinitely in all directions, remains boundless and hence defies identification. To argue that (the) universe can have a beginning or an end is to argue that the unidentifiable can paradoxically gain or lose an identity.

QUOTATIONS

1. The idea of something existing without ever having begun is so alien to our framework of reality as to be virtually incomprehensible.   Andrea J. Zwiebel (2)

2. Hannes Alfen who won the Nobel prize for physics in 1970, has argued that the Big Bang never happened — that the observational evidence is consistent with a universe that has existed forever, always changing and evolving, a universe without a beginning or end.
Eric J. Lerner (3)

3. First glimpses of the heavens by a new spacecraft looking for evidence of the beginning of existence have revealed an almost too perfect universe... . The craft..., so far has not detected any such disturbances. ...Dr. John Mather ...said..."But it's hard to understand that with all the large features we see in the universe, we have no evidence of their origin."   Warren E. Leary (4)

4. The universe seems to be organized in even more vast chains of galaxies...that defy understanding in terms of current theory of cosmology. ...Jeremiah P. Ostriker said..."...It could mean there's some vital missing ingredient in standard theories of development of the universe."   John Wilford Noble (5)

5. If Euclidean space-time stretches back to infinite imaginary time... : God may know how the universe began, but we cannot give any particular reason for thinking it began one way rather than another. ...One

could say: "The boundary condition of the universe is that it has no boundary." The universe would be completely self-contained and not affected by anything outside itself. It would neither be created nor destroyed. It would just BE.   Stephen Hawking (6)

6. ...Sir Fred Hoyle...who, five decades ago, derisively coined the term "Big Bang" for a cosmic theory which he did not and does not agree, makes plain...that the Big Bang is "a huge facade based on no real evidence at all."
                                        James Glanz (7)

7. If an explosive event like the Big Bang...happened once, ...it could happen an infinite number of times, and well may have. ...If there are parallel universes elsewhere, each would have started with its own Big Bang... . The romantics may have outdone themselves with their eternal inflation and a multiverse instead of a universe. ...Human beings...might find solace..., if there is anything to the anthropic principle in thinking that theirs is a defining presence in the one universe they know and are trying to comprehend.
                                  John Noble Wilford (8)

8. The Big Bang theory, for example, poses some rather obvious questions: Why did the Big Bang happen in the first place, and what, if anything, preceded it? Science itself suggests that we may never know, because the origin of the universe is too distant from us in space and time. ...Even the description of the physical world built up in the first three-quarters of this century has glaring gaps. The most obvious is that there is no way as yet of reconciling quantum mechanics with Einstein's theory of gravitation. That means there is no way of describing in detail the early stages of the Big Bang with which the universe is supposed to have begun... .John Horgan (9)

9. Scientists still have a way to go before they can calculate just how that inflating speck [the Singularity Point of

the Big Bang] popped into being; still they are close enough that various competing theories have set off a debate about whether it is proper to speak of a beginning of cosmic time, or whether the expression has no meaning. James Glanz (10)

10. We know that the stuff of the universe can both be conserved and change. ... Everyday experience is filled with stuff apparently coming into existence and then vanishing. Rain falls from the sky, forming lakes and rivers. In the dry season these can vanish, leaving no trace. The dramatic and manifold changes of the world suggested to some that things could spontaneously arise from nothing. Mark E. Eberhart (11)

REFERENCES

1. Wright, Robert (1988, Apr), Did the Universe Just Happen?, The Atlantic, p 29

2. Zwiebel, Andrea J. (1991, Jun 18), Letters to the Editor, New York Times, p A18

3. Lerner, Eric J. (1991, Jun 3), For Whom the Bang Tolls, New York Times, p C1

4. Leary, Warren E. (1990, Jan 14), Spacecraft Sees Few Traces of a Tumultuous Creation, New York Times

5. Wilford, John Noble (1987, Nov 10), Massive Clusters of Galaxies Defy Concepts of the Universe, New York Times, p C1

6. Hawking, Stephen W. (1988), *A Brief History of Time*, Bantam Books, New York, p 136

7. Glanz, James (1999, Sep 14), What Fuels Progress in Science? Sometimes a Feud, New York Times, p F1

8. Wilford, John Noble (1999, Feb 9), New Findings Help Balance the Cosmological Books, New York Times, p F1

9. Horgan, John & Maddox, John (1998, Nov 10), Resolved: Science is at an End. Or is it?, New York Times Book Review

10. Glanz, James (2001, Jan 28), Bang, You're Alive! On the Verge of Re-creating Creation. Then What?, New York Times, p 1wk

11. Eberhart, Mark E. (2003), _Why Things Break_, Harmony Books, New York, p 86

8

TIME, SPACE, AND UNIVERSE WITHOUT END

Time does fly.
Space does flow.
Whence do they come?
Whither do they go?
H.E.T.

What are time and space — really? In and of themselves they are nothing (in a literal sense — not a thing). But each serves as an important conceptual framework to give meaning to certain aspects of our life experiences. I recently saw a good definition of time as "nature's way of keeping everything from happening all at once," and space as "nature's way of keeping everything from happening to me."

The concept of space serves as a framework for the mental assignation of location of distributed masses and energy fields (e.g., light and heat) which we perceive as extending outwardly from ourselves to our immediate surroundings and by projection to places heard of, but unseen, and on into the starry heavens above, barely imaginable. Space acquires meaning for us as a result of the concurrent observation of different things and their locations/directions relative to ourselves and to each other. The concept of time serves as a framework for mentally managing several sequential phenomena. These include observed changes in location/ direction and/or quantity/ magnitude/ intensity of masses/energies which we can simply call activity.

The spatial and temporal natures of (the) universe have been the subject of a lot of publicity concerning major scientific construction projects; e.g., the Hubble telescope and the Super Collider have been offered to the public as instruments respectively capable of "seeing to the edge of the universe and beyond" and for "the re-creation of the very earliest fractions of a second after the Big Bang". As a professional engineer, I consider myself a scientifically oriented person and generally am supportive of scientific progress. However I certainly hope that there was considerably more to justify the tremendous expense of those projects than implied by such media hype.

I am utterly amazed that so many of our leading scientists subscribe to and support the "Big Bang" theory, as seen by references in article after article appearing in the scientific and general news media, and that there is virtually no reference and support for an alternative Steady State theory which is based on the acceptance of the reality that (the) universe was never created in any sense that is, or ever can be comprehensible.

I recall a story told to me by a good friend who grew up in Greece, about an old village patriarch lying on his death bed with his family and friends gathered around him. His son sat next to the bed, and after a prolonged silence the son suddenly said "Wait Father, don't go yet. Give us one last wonderful thought to remember you by." The old man lay still with his eyes closed, but slowly the whiskers began to quiver and the lips started to move. The son leaned close to his father and heard a barely audible whisper, "Life is like a well." The son's face lit up and he turned to repeat joyously to others in the room, "Life is like a well". This message was passed on out of the room to friends standing in the hallway and doorway, out into the courtyard and on down the road.

"Life is like a well" was repeated until it got to a beggar on the edge of town. When the beggar heard the message, he thought for a moment and asked, "What does it mean, 'Life is like a well'?" Now this question started back into town, up the road, into the courtyard, into the doorway, through the hallway and finally back to the son.

"What does it mean?" The son thought for a moment, shook his head and then spoke to his father again. "Wait Father, don't go yet. Tell us. What does it mean — 'Life is like a well'?" The old man still lay quietly with his eyes closed, but once again his whiskers began to quiver and his lips started to move. The son leaned close again to hear the bare whisper, "So? Life is not like a well."

Wouldn't it be refreshing if in a similar manner a few of our leading astrophysical scientists simply admitted that the idea of a point in time before which there was no time, or of a point in space beyond which there is or was no space is devoid of meaning. That an idea may be claimed to have meaning by an isolated few specialists is not enough. To have lasting meaning, the body of reasonably well educated "shepherds" of society must ultimately believe that they understand a scientific concept well enough to give it needed popular support.

I understand that adequate communication in this case is a very real problem. People often assign unique, personal meanings to words. (George Bernard Shaw once said that, The United States and Great Britain are two nations separated by a common language.) For example, the word universe means one thing to me, but obviously something else to others. To me, (the) universe is the infinite expanse of space and its content of masses and energy fields, which have existed forever from the past, and will continue to exist forever into the future in one form or another. When I read an article that refers to the creation or the extinction of "the universe," or the occasional article that theorizes about a multiplicity of "universes", I imagine confronting the authors and telling them, "Look, I don't know what you're talking about. Your 'universe' is obviously different from my universe. I'm talking about the one that existed before your 'universe' was created (i.e., recognized as a region within my universe) and will continue to exist after yours is extinguished (i.e., no longer recognizable in its present form) or the 'Mother of All Universes' that contains all the different kinds of 'universes' that anyone wants to talk about."

I see an example of this communication problem in the field of art, for instance where people talk about a "picture". An art critic may write an entire review about a"picture" hanging on the wall of some gallery. But consider for a moment what it is that he/she saw and wrote about. It likely was a frame with a piece of canvas having some oil colors spread around on it. All of these components contributed to form the "picture" the thing that the critic saw. But have you ever read a description of a frame in an art review? The picture that the reviewer describes is only the canvas and the oil colors. Thus on one level we think about a framed picture as the whole thing, but talk about the picture, which is only a part of the whole, on another level. In this case, we recognize that the frame is what we put it in and the picture, canvas and oil, is what you put into it.

I believe that (the) universe is considered in the same confusing manner. Most cosmologists seem to be describing a "universe" or "universes", which are only part of (the) universe. The space is what you put it in and the masses and energy fields are what you put into it. Together they comprise (the) universe. Space is just a place to put things, like a big empty closet, except that astronomical space has no walls, ceiling or floor anywhere. As such, it cannot have any boundary or property of its own (e.g., curvature or background radiation — properties actually attributed to the content of space), as do masses and energy fields, which have commonly known properties. Einstein's general theory of relativity does not describe the curvature of space in itself, but the curvature of spacetime.

To accept the idea that (the) universe is a concept of space extending to infinity in all directions from any real or imagined points, and including everything, mass/energy, distributed throughout that space, we must realize that it covers a lot of territory. To appreciate the extent of such a universe, it is helpful to understand the mathematical concept of the numerical value of infinity. Infinity is a term which describes an unreachable "never-never land" in quantification. It is not merely the largest number that anyone can imagine. A very young child, for example, just beginning to understand the idea of quantification must first recognize the difference between one and two identical objects before separating those from "more". Members of some primitive tribes only recognize a quantity up to 10, and any greater quantity is simply expressed as "many." Infinity is just such a quantity, beyond anyone's ability to imagine.

In that sense, imagine yourself as a random point of reference within (the) universe, with your arm outstretched horizontally, say in a (+) infinity direction. Now with a mighty stretch of imagination, picture a point, a planet called "A" as far away as possible on that line formed by your arm, skillions of light-years away. Planet "A" would then be at the farthest limit of comprehension of finite space and infinity would be beyond. But you need not stop there, because as a mental exercise its easy to imagine yourself instantly transported out to planet "A" and from there, still with your arm pointing in the same (+) infinity direction, you get another chance to imagine another way station an infinite distance farther beyond planet "A". And you can keep up this mental game as long as you like, but eventually you'll get tired of it and start thinking about something else, but rest assured that whatever way station in space you might have reached before mental exhaustion, the concept of an infinite distance would have allowed you to go farther toward a goal, but never reaching it.

In this fashion, (the) universe may be conceived as composed of two parts — one contained within the other. The first, smaller part of (the) universe is "our" universe — that part which can be observed, measured and comprehended in a real way. The scope of "our" universe varies from person to person, culture to culture, and from time to time. As the signal strength of our radio telescopes increase, "our" universe increases. Beyond that will always lie the "outback" second part, incomprehensible, but remaining component of (the) universe.

We believe that (the) universe is unlimited because no limit has ever been found. If we do not believe in an infinite universe, then we must address the question about what sort of boundary would limit (the) universe, and what sort of space, if any, would exist outside of that boundary — perhaps another universe of a different kind? If that was a contention, I would much prefer to include all different "kinds of universes" together in an infinite universe and refer to them instead as different regions of (the) universe.

Specifically, in regard to the longevity of (the) universe, while I believe that space and time (the framework) continues forever, masses and energy fields may come and go as recognizable objects or phenomena with time and according to Einstein's relationship, be interchangeable in form within any specific region of (the) universe.

Only after we accept the idea that time and space are without boundary conditions (i.e., infinite) can we move on to the next big question about the distribution of masses/energy fields throughout space, and how that distribution might change, if at all, with time.

One tenet about (the) universe, stated about 1920 by Alexander Freidmann, a Russian physicist and mathematician, is that any observer should see basically the same patterns and behavior of celestial bodies, looking not only in every direction from a single point as from Earth, but similarly from any other points in (the) universe. A static universe with more-or-less uniformly spaced bodies or clusters of bodies, with nothing really happening, would satisfy that condition. But "our universe", so far as we are able to see, does not seem to be quite that perfect. Recent reports from astronomers using the newest powerful telescopes confirm that the farther they are able to look out into space, the more uniform in any direction (the) universe appears to be. (1, 2, 3) But the activity of celestial bodies that has been seen by humans on Earth — a time span which compares to our reckoning of astronomical time like a single grain of sand to a beach — indicates that it is certainly not static. We have orbiting solar systems and comets, rotating binary stars, matter streaming toward black holes, swirling and merging galaxies, and supernovae to be seen in surprising frequency.

How then, does "our universe" satisfy the Freidmann condition? A purely expanding universe, which some scientists claim is the current mode of "our universe", or a purely contracting universe, which some claim to be its fate, would not do so. Anyone watching a 4th of July fireworks display knows that an expanding system, a uniformly and symmetrically expanding system, looks the same in every direction only from a single point at the center of the system. The same is true for a contracting system.

Apparently there is some disagreement between leading astronomers as to whether our region of (the) universe is actually expanding or contracting. In this regard, Carl Sagan says, "...something very strange is going on in the depths of space." (4) Alan Dressler reports in a study of large scale streaming of galaxies, that many theorists feel that the mass density of (the) universe is exactly what is needed to arrest cosmological expansion in the distant future, even though observation of some galaxies do not support that idea. He states that "...[his] study and other attempts to measure the density of matter on the largest scales, however, point to an overall density close to the critical value, and perhaps to a universe in perfect balance." (5) Frankly, I consider the possibility that our region of (the) universe may be expanding as part of an oscillatory mode of changing from expansion to contraction, and further that (the) universe might simultaneously contain some regions that are expanding and some other regions that are contracting, as healthy signs of overall systemic balance as the ebb and flow of the ocean tides.

A universe full of random, chaotic activity would satisfy the Freidmann condition if it were understood that observations had to be made and averaged over extended periods of time. Imagine sitting at the sea shore and watching the waves. If you watched only for a few hours while the tide was going out, you would have one impression of the water's behavior; or if you happened to watch only for a few hours while the tide was coming in, an opposite impression. Over an extended period of time, wave and tidal motions look pretty much alike at all of the beaches, and the water level averages out to a mean level.

I think that the Freidmann condition is reasonable and that the latter model of chaotic motion of bodies throughout (the) universe is the best choice. Chaotic rearrangement is apparently going on within our own Milky Way galaxy and we have no reason to believe it is different in that respect from any other galaxy.

If cosmological evidence indicates that a big bang was involved in the creation (re-creation) of our region of (the) universe then I can accept and support that notion, since, as I will discuss later, I believe that big bangs at random points and at random times throughout (the) universe are a part of the natural scheme in maintaining a balance of things. But to ascribe the "creation" of (the) (whole infinite) universe to a single big bang is like being one of the six blind men of Indostan in John Saxe's wonderful children's poem, who each think that the part of the elephant that he is holding characterizes the entire animal.

While the "Big Bang" concept of the "beginning of the universe" involves a vision of untold amounts of matter being flung outwardly, that is, expanding into space from a singularity point, the "end of the universe" is pictured by some as the exact opposite effect, like a movie film run backwards, with matter being pulled back together, contracting by gravitational attraction. There is a sophomoric physics joke in which a professor tells his students that he's calculated that the entire universe will collapse into one big ball in some 3 billion years. A half-awake student in the back of the room shoots up his hand and asks the professor to repeat his remark. The professor obliges. The student breathes a sigh of relief and exclaims, Thank God! I thought you said 3 <u>million</u> years.

After mulling over this joke for some time, I concluded that even 3 billion years may be too soon to believe that (the) universe would collapse. More likely, it never will. Why should we believe that we exist in a universe that is due to totally self-destruct at some time in the future? I prefer to believe that there is a timeless overall stability in the nature of (the) universe, while leaving plenty of room for local and even galactic scale disturbances.

The really big question is how (the) universe which we can see at this point in time to have essentially uniformly distributed masses and energy fields in locally chaotic motions, might stay that way over astronomical periods of time and ultimately forever. This question contradicts our prevailing concept of attractive gravitational force, the predominating shaper of celestial structure, which leads us to expect that (the) universe's ultimate fate is to end up as one great big, cool ball, regardless of the process that it takes to get there.

One possible argument to support the notion that all of the bodies in (the) universe might <u>not</u> end up in one big ball could be fashioned from the concept of "chaotic clay" without doing violence to our present understanding of gravitational attraction between masses, as follows.

Imagine that the earth (1st order attractor) is approaching the time when the last few of its regularly orbiting bodies — moon, satellites, space junk, (forget about comets) — are on the verge of re-entry. But just before that happens, the earth is drawn close enough to the sun (2nd order attractor), so that the sun's gravitational force is capable of disturbing what remains of the earth's orbital system and causes that system to start streaming toward the sun. Then, just before the last few of the sun's orbiting bodies are pulled into its atmosphere, another larger body, like a small neutron star or black hole in the Milky Way (3rd order attractor), begins to exert its influence on the remains of our solar system and causes them to start streaming toward it.

I could carry this illustration on further, but I hope it is unnecessary. Using this scenario, we can in turn imagine discrete parts of (the) universe constantly contracting through gravitational attraction, but as an "entire" universe, never quite making it, because in an infinite universe we can always imagine a next higher order attractor that could emerge and interfere to defeat the completion of a local lower order gravitational accretion.

Given the progression of the scale of attractors required to support this scenario, I admit that such an argument is unrealistic. Not just because it is so improbable, but because there is more required of a mechanism that surely must now be in operation to keep (the) universe from ending up in one big ball. The anthropic view dictates that if (the) universe could end up in one big ball, it would have happened long ago and we would not be here to wonder about it now.

There must be a mechanism that keeps (the) universe forever as we now see it, a space distributed and filled more or less uniformly with a full range of particles/bodies, from singular atoms floating in the space vacuum to the humongous black holes. The mechanism must involve not only a process to limit the growth, the gravitational accretion process of making big ones out of little ones to a maximum allowance, but also a process of regeneration, reducing big ones back into little ones.

The process of reducing big ones back to little ones is especially challenging to our thinking because of the unimaginable range in changing scale factor required from start to finish. But it should not really be a surprise since the reduction of material from black hole scale back to subatomic particle scale is merely the opposite of the building up process.

Some red giant stars that have exhausted their supply of hydrogen are estimated to be so big that if one of them were in the position of our sun, it would envelop our solar system to beyond the earth's orbit. That is a class of star that can be seen. How much larger might the largest black holes be? It is believed that after a supergiant star uses up its supply of hydrogen fuel in its nuclear reacting innards, it begins to collapse. As it collapses, its internal temperature increases and reactions between atomic nuclei become more rapid. As the contraction progresses the remaining material becomes more highly compressed and the star's speed of rotation increases. Fred Hoyle estimates that for such a collapsing star, just before the outburst of a supernova, the core temperature might be about 15×10^8 C, the density of core material might be about 2×10^{11} lb/in^3, and its surface speed of rotation about 1×10^8 mph. He also estimates the violence of explosion of a supernova as equal to about 1×10^{24} hydrogen bombs all going off at the same time. (6)

Hydrogen is not only the lightest, but the most abundant element in (the) universe and through its reaction to form the next heavier element, helium, it serves as the primary fuel to generate the thermal energy within celestial bodies. Fred Hoyle argues that in order for (the) universe to be infinitely old, there must be a mechanism to continually replace the hydrogen being consumed in all of the stars. Otherwise, the supply of hydrogen in an infinitely old universe would be completely exhausted. Therefore, our universe, which consists almost entirely of hydrogen, must either be very young (from an astronomically recent time of "creation") or if it is infinitely old, it must have a mechanism for re-creation of hydrogen. But he says hydrogen cannot be produced in any appreciable quantities through the breakdown of other elements.

Hoyle, along with H. Bondi and T. Gold, fellow Cambridge scientists, describes a process that they call continuous creation — in contrast to the "Big Bang" instantaneous creation of all matter in (the) universe. Hoyle says, "On scientific grounds this big bang assumption is much the less palatable of the two. ...Continuous creation...can be represented by precise mathematical equations whose consequences can be worked out and compared with observation. On philosophical grounds... I cannot see any good reason for preferring the big bang idea... since it puts the basic assumption out of sight where it can never be challenged by direct appeal to observation". (6)

While Hoyle does not say this exactly, I believe that the new material is formed in the continuous creation process not by the chemical or nuclear reaction of other (heavier) elements, but by the slow process of recombination of subatomic particles, liberated into space from the cataclysmic stellar explosions. Hoyle says, "Now if the collapse proceeds far enough before the rotary forces break up such a star, these nuclear reactions, must start to absorb energy instead of generating it. This situation, which goes the opposite way of everything we have considered so far, is due to the large scale production of free neutrons." (6) All it takes to create a new hydrogen atom is one free proton combining with one free electron. Hoyle says, "The new material does not appear in a concentrated form in small localized regions but is spread throughout the whole of space. The average rate of appearance of matter amounts to no more than the creation of one atom in the course of about a year in a volume equal to that of a moderate size skyscraper." (6)

My search for an explanation of how (the) universe might not be driven toward the destiny implied by the conventional understanding of gravitational attraction leads me down an unconventional path. Like the man in Robert Frost's famous poem, "The Road Not Taken", I too have chosen to take the one less traveled by, and that has made all the difference. (7)

QUOTATIONS

1. A team of 50 astronomers and data processors [have]... concluded that...the universe looks largely the same in any direction.   John Noble Wilford (1)

2. ...[O]n the largest scale, the arrangement of matter in the universe is uniform, on an intermediate level, galaxies are found in gravitationally bound aggregates referred to as 'groups' or 'clusters'. The gravitational potential which binds galaxies within a cluster often also binds a vast cloud of hot gas which fills the space between and around the galaxies.   Philip A. Charles (2)

3. Jaan Einasto says, ...the analysis of data on galaxy clusters indicates that the large-scale structure of the cosmos is an orderly rectangular, three-dimensional latticework of clusters and voids. The crisscrossing lines of clustered matter appear to be spaced at more or less regular intervals of 391 million light-years. If their findings hold up, the Einasto group suggests, scientists may be forced to look for some hitherto unknown process that imposed an order on the universe.   Kathy Sawyer (3)

4. Categories such as time, space, and number represent the most general relations that exist between things; surpassing all our other ideas in extension, they dominate all the details of our intellectual life. If humankind did not agree upon the essential ideas at every moment, if they did not have the same conception of time, space, cause, and number, all contact between their minds would be impossible... Emile Durkheim, "Les formes elementaires de la vie religieuse." (Paris, 1912), pp 22-23 (8)

5. Einstein abolished the concept of space as a container of the material universe. Einstein's space is a <u>constituent</u> of it.   Reginald Kapp (9)

6. ...[N]o point in the universe can be special; therefore, the universe cannot have a boundary.

Stephen Hawking (10)

7. Relativistic cosmology in general and standard models in particular have the curious and unsatisfactory feature of a spacetime singularity. The appearances of infinities is considered disastrous in any physical theory. In general relativity it is worse, since the singularity refers to the spacetime structure and physical content of the universe itself.   Jayant V. Narlikar (11)

8. ...[S]uperclusters, agglomerations of galaxies, and galaxy clusters...stretch across millions of light years of space. ...Dr. Melott said..."We know structure like the ones seen can be produced by gravity, but we've had no evidence of the particular kinds of flows that bring them about." ... "Superclusters are so big their basic pattern could not have changed much..." Dr. Henry said. "We are probably seeing the original pattern of structure in the universe."   John Noble Wilford (12)

9. ...[T]here are so many galaxies in the universe that, somewhere in the sky, supernovae bright enough to study erupt every few seconds.   Craig J. Hogan et al (13)

10. Interestingly, the idea that the universe is homogeneous dates back to the sixteenth century. The Italian philosopher and scientist Giordano Bruno asserts in his book, _On the Infinite Universe and Worlds_, published in 1584, that the universe has the same structure throughout, consisting essentially of an infinite repetition of solar systems (in his words, ... "assemblies of worlds").   Mario Livio (14)

11. ...[A] major puzzle in cosmology: the universe's large scale homogeneity. The need to explain the uniformity led to the theory of inflation, but inflation has run into difficulty of late, because in its standard form it would have made the cosmic geometry Euclidean — in

apparent contradiction with the observed matter density. This conundrum has driven theorists to postulate hidden forms of energy and modifications to inflation.

Jean-Pierre Luminet (15)

12. Dr. Robin E. Williams ...said: "You can see a myriad of galaxies [with the Hubble Space Telescope]. There are large ones and small ones, red ones and blue ones, very structured ones and very amorphous ones. Most of the galaxies were never seen before Hubble. But we don't know the significance of all this yet. ...There is no evidence that we are seeing back to the point of first galaxy formation... ." ...This more expansive cosmic view is a far cry from that in the beginning of the century when astronomers assumed the Milky Way was all there was, the universe entire. ...The universe, statistically, looks largely the same in all directions.

John Noble Wilford (16)

13. Max Planck, the originator of quantum theory, ...proclaimed that the old ideas in science die only with those who hold them. ...Scientists may have to concede that even in this time of breathtaking astronomical discoveries, some things like the age of the universe remain beyond human understanding.

John Noble Wilford (17)

14. Dr. James E. Peebles [says about the "end of the universe"], ... "How can you resist speculating on the future of the universe. ...We have given up on the notion that the laws of physics are known absolutely. So it's awfully hard to know if they are the real truth or be sure that they won't fall sometime in the distant future, leading to a completely different outcome."

John Noble Wilford (18)

15. Why is the description of the universe always confined to what scientists say started with a Big Bang 15 billion years ago...? If the universe is infinite and eternal — isn't it? — Big bangs probably happen all the time, perhaps in infinite numbers. It seems absurd to think that all the matter, gases and energy in the infinite universe reside solely in the pocket of the realm where we reside. Maybe the galaxies at the edge of our universe are moving faster away because of the gravitational pull of other universes.

Duncan A. MacDonald (19)

16. ...I believe we can and should try to understand the universe. ...We don't yet have a complete picture, but this may not be far off. The most obvious thing about space is that it goes on and on and on. ... As far as we can tell, the universe goes on in space forever. Although the universe seems to be much the same at each position in space, it is definitely changing in time.

Stephen Hawking (20)

17. Recent loop quantum gravity calculations by Martin Bojowald of the Max Planck Institute for Gravitational Physics in Golm, Germany, indicate that the big bang is actually a big bounce; before the bounce the universe was rapidly contracting. ... It is not impossible that in our lifetime we could see evidence of the time before the big bang.

Lee Smolin (21)

18. So, when did time begin? Science does not have a conclusive answer yet, but at least two potentially testable theories plausibly hold that the universe — and therefore time — existed well before the Big Bang. If either scenario is right, the cosmos has always been in existence and, even if it re-collapses one day, will never end.

Gabriele Veneziano (22)

REFERENCES

1. Wilford, John Noble (1998, Nov 24), A New Baby Picture of the Southern Sky, New York Times, p F5

2. Charles, Philip A. & Seward, Frederick D. (1995), *Exploring The X-ray Universe*, Cambridge University Press, Cambridge (U.K.), p 310

3. Sawyer, Kathy (1997, Jan 13), Astronomy: Seeing Order in Cosmos, Washington Post, p A02

4. Sagan, Carl (1980), *Cosmos*, Random House, New York

5. Dressler, Alan (1987, Sep), The Large Scale Streaming of Galaxies, Scientific American, p 46

6. Hoyle, Fred (1950), *The Nature of The Universe*, Harper & Brothers, New York, pp 82, 124, 125

7. Frost, Robert (1953), *Robert Frost's Poems*, Pocket Books, Inc., New York, p 223

8. Tufts, Edward (1991), *Visual Explanations*, Graphics Press

9. Kapp, Reginald O. (1960), *Towards a Unified Cosmology*, Scientific Book Guild, London, p 148

10. Hawking, Stephen W. (1988), *A Brief History of Time*, Bantam Books, New York, p 40

11. Narlikar, Jayant V. (1983), *Introduction to Cosmology*, Jones & Bartlett Publishers Inc., Boston, p 394

12. Wilford, John Noble (1999, Jan 26), Superclusters of Galaxies Shed New Light on Cosmic Architecture, New York Times, p F1

13. Hogan, Craig J. et al (1999, Jan), Surveying Space-time with Supernovae, Scientific American, p 46

14. Livio, Mario (2000), *The Accelerating Universe*, John Wiley & Sons, p 44

15. Luminet, Jean-Pierre et al (1999, Apr), Scientific American, p 90

16. Wilford, John Noble (1996, Jan 16), Suddenly, Universe Gains 40 Billion More Galaxies, New York Times, p A1

17. Wilford, John Noble (1995, Jan 10), New Measurements on Age of Universe, New York Times, p C10

18. Wilford, John Noble (1997, Jan 16), At Other End of "Big Bang" May Simply be Big Whimper, New Yo rk Times, p A1

19. MacDonald, Duncan A. (2002, Jan 8), (Letter to the Editor), New York Times, p F3

20. Hawking, Stephen (2001), *The Universe in a Nutshell*, Bantam Books, New York, pp 69,71

21. Smolin, Lee (2004,Jan), Atoms of Space and Time, Scientific American, p 75

22. Veneziano, Gabriele (2004, May), The Beginning of Time (the myth of), Scientific American, p 65

9

(THE) UNIVERSE AS A PERPETUAL MOTION MACHINE

Some scientists, with whom I do not agree, argue that if the laws of thermodynamics are valid, (the) universe cannot last indefinitely as it appears to be now. They say it is in the process of dying. I argue that (the) universe is alive and well and will continue to exist forever as it has in the past since it alone possesses the qualification to act as a perpetual motion machine. I readily concede that localized changes are occurring within (the) universe, thus failing to maintain a historically recognizable form for an individual observer. But conditions throughout (the) universe, on average, should always look the same.

Simply stated, it takes a lot of energy to keep (the) universe running - producing heat, light and motion. The prevailing thought is that a limited supply of energy is available to be expended within (the) universe. Energy to keep (the) universe running is supplied or more accurately, transferred primarily in the form of heat that flows of its own accord from bodies at a higher temperature (like our sun) to bodies at a lower temperature (like our planet Earth) and also to the ultimate heat sink, the celestial background of space ("black void"), which is currently estimated to be at 2.7 K absolute temperature scale. In the absence of any self-acting restorative process, the end result of transferring a limited supply of heat from the hot bodies to the cold bodies would be a cooling of all the hot bodies and related warming of all the cold bodies.

That would leave everything as a mixture of uniformly lukewarm "soup" with no more temperature differences between bodies to facilitate the normal transfer of heat energy. This view of universal finality, which some scientists believe, is neither as dramatic as that of (the) universe collapsing from an overabundance of attractive gravitational force into a singularity, nor expanding indefinitely from an insufficiently attractive gravitational force toward a state of zero density. It is one which raises a serious question about how (the) universe might manage to operate forever in essentially the same condition as we see it now. For (the) universe to have any self-acting process of restoring thermal energy to a body (i.e., raising its temperature again after it has cooled), would put it in the class of a perpetual motion machine — something supposedly forbidden by one of the laws of thermodynamics.

I argue that that specific law of thermodynamics which seems to be valid in its application to our work-a-day-world and throughout distinct regions within (the) universe, does not apply in the same way to (the) (infinite) universe. The first law of thermodynamics basically states that the sum total amount of energy in all forms (e.g. nuclear, electrical, chemical, or mechanical) is constant within any "closed system", and by extension, within (the) universe. Within any recognized system, defined by a boundary and initially containing discrete amounts of energy in all the various forms, the total amount of energy must remain unchanged. Additional amounts of it cannot come from nowhere, and lessening amounts cannot disappear to nowhere. Changes in the amount of any specific form of energy result from transformation of one type to an equivalent amount of another. This may involve natural or manmade processes/ mechanisms, (e.g.) solar heat creation of wind, or evaporation of water into Earth's atmosphere, chemical energy to electrical energy as within a battery, mechanical energy to electrical energy as within a generator, or electrical energy to thermal energy (heat) as within a toaster. Also in accordance with Einstein's well known equation,

$E=Mc^2$, mass can be converted into energy and vice versa in the right nuclear environment.

Note that in any real "closed system", whether the boundary be arbitrarily drawn around an atomic-level structure or around a galactic-scale structure, the reservoir of energy within is considered constant and finite. (The) universe may be similarly considered to have a reservoir of all the different kinds of energy that is constant. But the total amount of energy in the reservoir would have to be infinite since the reservoir of space is infinite, and that makes the amount meaningless. Infinity plus or minus any amount is still infinity.

The second law of thermodynamics basically states that whenever energy in one form is unnaturally forced to be changed to another form through some artificial (e.g., manmade) machine or process toward the production of "useful work", there will be some loss (in the equivalent amount) of energy during the conversion process. Imagine for example that we have a "Rube Goldberg" system with a weight hanging on a cord which is wrapped around a pulley whose shaft drives an electrical generator as the weight falls from an initial height X. Also imagine that the generator charges a battery that is used to power an electric motor connected to the pulley shaft that, when required, turns backwards and picks the weight up again. In this system, we would always find at the end of such a cycle that the weight can never be returned completely to its original height X, no matter how carefully the components in the system are made. The ultimate reason that a portion of the system's initial potential energy is unavailable to be converted to useful work (i.e., returning the weight to its original height) derives from the fact that no machine or process can be made totally free from the effect of friction which generates heat and noise. Additionally, no "closed system" operating at a temperature level above its surrounding environment (2.7 K for the "black void"), can be devised with a blanket of thermal insulation so perfect as to have zero heat loss. Note also that, if there were any diversion of generated electricity to do other useful work outside the "closed system", it would further increase a final energy deficit. To return the system to its original energy level (height of weight) at the end of each cycle, visualize a form of energy input — a hand having to reach into the "closed system" and lift the weight as needed, after the battery-driven motor runs out of power, the rest of the way up to the initial height. Thus we see that the amount of energy that must be added to a "closed system" is equal to the amount lost in the form of irrecoverable heat to the environment surrounding the system plus any other amount transferred out of the system.

The laws of thermodynamics are based on physical observation and not subject to mathematical proof. They are treated in mathematics as idealized models of "closed systems" and are accepted as valid because no exception has ever been found (within (the) universe). Thus the second law not only dictates that you can never have a "closed system" that puts out more energy in the form of useful work than is put into it (a perpetual motion machine of the 1st class), but also dictates that you can never have a "closed system" that will just keep itself running forever with no energy input, even though it may not put out any "useful work" (a perpetual motion machine of the 2nd class). Some extremely clever craftsmen have offered serious challenges to this prohibition by building nearly frictionless clocks with ingeniously hidden "motors." None have been known to run "forever" without the input of some energy, even if nothing more than changes in atmospheric illumination, temperature or pressure. I think that most people who should be familiar with this law have lost sight of an important fact: that the prohibition against a perpetual motion machine of the 2nd class is conditioned by the unavoidable loss of heat energy from closed systems located within (the) universe to a surrounding environment. But (the) universe in its infiniteness has no boundary and hence no surrounding environment to irreversibly absorb any heat energy losses. In a strange way, this makes it the only system that can truly be said to be closed (i.e. with <u>nothing</u> crossing its boundary) because it has no boundary. Additionally, energy loss from any region within (the) universe is simply, and totally, an energy gain to adjacent regions. Thus (the) universe can be and arguably is the only possible perpetual motion machine.

My version of a likely scenario for the perpetual recycling of energy and mass throughout (the) universe would be the following: A specific body randomly located in some region of (the) universe grows by attractive gravitational accretion of nearby lesser bodies, bringing together materials that react kinetically and/or chemically to raise the temperature of the body. It finally reaches such an enormous state of mass and energy content (e.g., neutron star or black hole) that it vaporizes and/or explodes material in all directions (a big bang), thereby losing material and energy to adjacent regions. The cooled remnant body of the explosion, which may or may not retain its most recent gravitational identity, then floats around in space. Eventually it attracts other lesser bodies or is attracted to some greater body of the same gravitational identity. This convergence again brings together materials that react kinetically and/or chemically to raise the temperature of the body...ad infinitum.

Here I suggest that the result of the big bang is not only a redistribution of energy and mass from a region of (the) universe where its principal inhabitant has reached a maximum allowance in mass/energy content, but enables a seeding for re-growth of bodies in the region which may have a different gravitational identity, leading eventually to another galaxy whose bodies are mutually attractive to each other within the galaxy, but may not be attractive to its nearest galactic neighbor.

This scenario suggests that a region of (the) universe will cyclically give up mass and energy to adjacent regions and later get them back again. Reginald Kapp says that galaxies go through cycles of about 3.5 billion years of dwindling (losing material and energy) before gaining again - thereby remaining stable (and not growing to infinite size). (1)

QUOTATIONS

1. Ludwig Boltzmann, in 1877, propounded a new concept — that the universe as a whole must like any closed system, tend toward an equilibrious state of entropy: it will be completely homogeneous, the same temperature everywhere, the stars will cool, their life-giving energy flow will cease. The universe will suffer a "heat death". Any closed system must go from an ordered to a less ordered state — the opposite of progress.　　　Eric Lerner (2)

2. ...[T]heorists have painted a vivid picture of the heat death of the universe, after all the stars that have been made or will ever be made have burned out. Since the dawn of civilization, mystics and prophets have speculated about the end of the universe, yet we are still here... . We are emboldened to ask questions about things we may never see directly because we can ask them. Our children, or their children, will one day answer them. ...We are endowed with imagination, and the history of science has taught us that our vision is limited only by our willingness to look. ...We can ponder whether any surviving intelligent life a trillion trillion trillion years from now would be content to let a dying universe die.　　　Lawrence Krauss (3)

3. The intense gravity of black holes is what makes them such efficient engines. ...Objects are pulled toward the horizon at a ...high speed, and en route they may collide with other objects and shatter. The effect is to heat the material near the hole, because the objects are moving at close to the speed of light, the kinetic energy available for transformation into heat is comparable to the energy inherent in the mass at rest ($E=Mc^2$). ...In this sense, black holes transform rest mass into thermal energy.　　　Jean-Pierre Lasota (4)

4. Attempting to investigate the microscopic properties of black holes, the gravitational traps from which not even light can escape, Dr. [Stephen W.] Hawking discovered to his disbelief that they could leak energy and particles into space, and even explode in a fountain of high-energy sparks.   Dennis Overbye (5)

5. The big bang [i.e., "Big Bang"] violates all conservation laws of physics, such as conservation of matter and energy.   J. V. Narlikar (6)

6. The second law [of thermodynamics] tells us that whenever energy is put to use it is degraded... . ...The energy of the wasted heat is still somewhere in the environment, in the form of randomly vibrating molecules...[and]...it can never be recovered. If it weren't for this loss, we could ...have a perpetual motion machine.   George Johnson (7)

7. ...[T]he universe, by definition, has no environment. There is nothing outside it.   George Johnson (7)

8. [J. B. S.] Haldane imagined a far future when the stars have darkened and space is mainly filled with cold, thin gas. Nevertheless, if we wait long enough statistical fluctuations in the density of this gas will occur. Over immense periods of time the fluctuations will be sufficient to reconstitute a Universe something like our own. If the Universe is infinitely old, there will be an infinite number of such reconstitutions, Haldane pointed out.   Carl Sagan (8)

9. "It seems that almost as soon as nature builds spiral galaxies into clusters," Dr. [Alan] Dressler said, "it begins tearing them apart."   John Noble Wilford (9)

10. It's just possible that quasars don't actually die, but simply become dim when they run out of matter to consume, and rekindle when new matter comes their way. On the other hand, this object in NGC6240 may consist of dark

110

matter in some entirely different form.

Joss Bland-Hawthorn (10)

11. "Blue stragglers have been an astrophysical puzzle for a long, long time, and it's on the way to being solved," said Dr. Allan R. Sandage... . In the densest galaxy clusters ...close encounters... lead to physical collisions causing two or more objects to coalesce. "...These collisions range from fender benders to head-on smashes," said Dr. Michael Shara... . "They touch off all kinds of mixing and thrashing motion, and soon it's as if the star is reborn. That's what I think we're seeing — 15 billion year old stars that are reborn with a luminosity typical of a 1 billion year old."

John Noble Wilford (11)

12. If the universe is to go on evolving for all eternity without running down to some final state, it must be continually rewound.   Paul Davies (12)

13. If the universe is evolving over an infinite span of time and if evolution toward greater complexity and higher energy flows is explicable in terms of natural processes, then the creation out of chaos is comprehensible.

Eric Lerner (2)

REFERENCES

1. Kapp, Reginald O. (1960), *Towards a Unified Cosmology*, Scientific Book Guild, London, p 161

2. Lerner, Eric (1991), *The Big Bang Never Happened*, Times Book Div of Random House, New York, pp 119, 7, 397

3. Krauss, Lawrence (1997, Jan 22), Not With a Bang or a Whimper, New York Times, p A21

4. Lasota, Jean-Pierre (1999, May), Unmasking Black Holes, Scientific American, p 40

5. Overbye, Dennis (2002, Jan 22), Hawking's Breakthrough Is Still an Enigma, New York Times, p F1

6. Narlikar, J. v. (1983), *Introduction to Cosmology*, Jones & Bartlett Publishers Inc., Boston, p 394

7. Johnson, George (1996), *Fire in the Mind*, Vintage Books, New York, pp 115 & 168

8. Sagan, Carl (1996), *The Demon-Haunted World*, Ballantine Books, New York, p 206

9. Wilford, John Noble (1994, Dec 7), Hubble Captures Photos of Universe in its Infancy, New York Times, p B9

10. Browne, Malcolm W. (1991, Apr 9), Astronomical Mystery: Dark Lurking Object, New York Times, p C10

11. Wilford, John Noble (1991, Aug 27), Cannibal Stars Find a Fountain of Youth, New York Times, p C1

12. Davies, Paul (1991, Jul 28), 20 Billion Years Isn't Forever, New York Times Book Review, p 20

10

MYSTERIOUS MATTERS

Whenever I stop to think about it, which isn't very often, I am truly amazed at the phenomenon of gravity. It is so basic to our existence and so taken for granted. Without it, of course, we simply wouldn't be here. It's always with us and we can't do anything about it, like the weather which everyone talks about. In New England, a common reply to anyone's complaint about the weather is that they should wait until tomorrow, and it will change. But no one except a few dedicated scientists and teachers seems to be talking about gravity. Which is understandable because generally it's a very dull subject. I can't imagine greeting my friends with "Well, what do you think gravity is going to do today?", or "Boy, we've sure had a good pull of gravity these last few days". However, if the subject should come up, and someone complains to me about the current state of gravity, I am now ready to reply that they should wait a few billion years and it, too, will change. The nature of gravity and how it might change, however, remain as mysterious as life itself.

Life and the various forms of energy inherent in matter have been, and will continue to be studied in more and more detail until perhaps one day someone will be able to identify the key ingredient which puts life into otherwise inert matter. Matter has been studied to the point at which subatomic particles and new "exotic" elements (e.g., Lawrencium and Einsteinium) are being discovered purely from the basis of mathematical prediction and intuitive hunches. Nothing has been found so far to indicate what the key ingredient is that adds the property of gravitational attraction to the basic inertial mass of discrete bodies. Weight — a product of gravitational attraction — is not a property that can be observed in an isolated body, like measuring the luminosity and radiation spectrum of a distant star. Weight is observed only as the active force of "natural attraction", which we call "gravity", between a pair of bodies ($F=GM_1M_2/d^2$). But yet, the property of gravitational force potential is inherent in an isolated mass, just as we understand that a single magnet should retain its potential to attract other magnets even if it were to be floating alone in distant space. The property of inertial mass of a discrete body is best observed by noting its reaction to any one of a variety of imposable forces (e.g., mechanical, gravitational, magnetic, radiation pressure, etc) which would have the capacity to change the body's position or velocity ($F=Ma$).

The phenomenon of gravitational force is just one of four forces known to be ingrained in the behavior of matter. The other three are: electromagnetism, which creates light, radio waves, microwaves, and other forms of electromagnetic radiation; the strong force, which binds neutrons and protons together in the nucleus of an atom; and the weak force, which makes atoms break in radioactive decay.

In 1928 Dr. Paul A. M. Dirac predicted the existence of a new set of particles which came to be known as antimatter, on the basis of his strong belief that there should be a symmetrical balance in mathematics and nature. In 1932 Dr. Carl Anderson discovered the antielectron (or positron) amid particle tracks that curved the wrong way in a cloud chamber. Later experiments with atom smashers capable of creating extremely high temperatures and pressures, produced antiprotons as well as antielectrons. Basically, the difference between electrons and antielectrons, as well as between protons and antiprotons, is simply a reversal of the positive and negative electrical charges. Otherwise, the antimatter particles have the same mass and other measurable properties as their ordinary counterparts of matter. But due to the electrical difference, matter and antimatter are not compatible. Antimatter particles created on Earth last only a few millionths of a second before their annihilation in a collision with matter. The expected behavior of antimatter in the gravitational field of ordinary matter has been uncertain — whether it would fall up or fall down on Earth. In 1996 it was reported that some atoms of antimatter had been created for 40 billionths of a second at the European Laboratory for Particle Physics (CERN). (1) An earlier report predicted that the antiproton particle would fall down with an acceleration 14 percent greater than that of ordinary matter, based on theoretical considerations, and an experiment was planned to confirm that. (2)

Dr. Robert L. Forward said: "The real question in everybody's mind is why the universe seems to be made up of matter when, on a cosmic scale, antimatter is just as easy to make. It's one of the outstanding mysteries." Dr. Floyd W. Stecker said: "An antigalaxy would look just like any other galaxy. For the universe as a whole it could turn out that there's just as much antimatter as matter." (3) While there appears to be hope for finding a great quantity of antimatter in space, it does not appear to be a good candidate for the mysterious matter which must counterbalance the gravitational attraction of ordinary matter and keep (the) universe from collapsing.

During the 1920s-30s, the idea of the Hubble expansion took root in cosmological thinking, along with the idea of the "Big Bang" creation of the universe. That theory has prevailed in spite of an inability to account for any more than about one to ten percent of the mass of visible celestial bodies needed to balance the dynamic forces believed to be involved in the observed motions, based on the theory of gravity as established by Einstein in 1917. Einstein first formulated his conception of a static, finite universe after developing the general theory of relativity, but later grew concerned that a static, closed universe could never remain static because it would collapse (contract into one big ball) from the everlasting prevalence of attractive gravitational force. Edgar Allan Poe had pointed out much earlier that rotation is what stabilizes galaxies and solar systems and keeps them from collapse. Einstein reasoned that something like rotation prevents the collapse of (the) universe, but clearly not rotation itself, since (the) universe could not be rotating relative to something else. He revised his equations of gravity by including a term called "the cosmological constant" to provide a repulsive force as a property of space, not of mass, which increased in proportion to the distance between bodies. He then set the value of the constant to provide for the theoretically-perfect balancing of gravitational forces to maintain a static, non-collapsing universe. (4)

During 1925-1927, Georges-Henri Lamaitre, after hearing about Hubble's expansion observations, studied Einstein's equations of gravity and found that the solution proposed by Einstein was unstable. A slight expansion between bodies caused the repulsive force to increase and expansion grew unlimited; a slight contraction led to total collapse. Different solutions of Einstein's equations and/or assignment of different values to his cosmological constant yielded theoretical models of universes that were expanding or contracting or statically balanced (though unstably — picture a pencil delicately balanced on its point and protected from the least vibration or wind), of finite or infinite age, and finite or infinite spatial extent. Every possible solution, though is limited in some way, either by an origin in time or by being closed in space, or both. In any event, a repulsive force of unknown origin was believed to be counterbalancing gravity, and according to the interpretation of Hubble's data, causing (the) universe to expand. (4)

Einstein later expressed regret at having created the cosmological constant, saying it was "the greatest blunder of my life". But some cosmologists still believe that Einstein was right in conceptualizing an antigravity force built into space itself that may not only be causing the Hubble expansion, but must, in the long run, keep (the) universe from collapsing. (5)

At the same time, however, cosmologists have developed the view that aside from whatever is causing the Hubble expansion observation, if our region of (the) universe is, in fact, expanding, there simply are not enough detectable bodies to provide the gravitational attraction force (glue) needed to account for the "togetherness" of visible galactic structures and the maintenance of their "interconnectedness", at the accepted Hubble expansion rate of 50-100 km/sec/10^6 pc. Viewed from this perspective, 90-99 percent of the mass needed to balance the theoretical equation of force and motion is missing. As a result, the term Missing Mass was adopted to describe the mysteriously elusive amount of ordinary matter which would have to exist throughout space, to provide the additional attractive gravitational force needed to keep the "observed expansion" in check. The elusive material is also sometimes called Dark Matter — not to be confused with Dark Energy, which is a proposed antigravitational force of space. Dark Matter is believed to exist in super-dense dwarf galaxies, too weak to emit any radiation, and/or in superfine particles forming halos around galaxies (called MACHOS — massive compact halo objects) which are not just dark but invisible. These particles are believed to be made of exotic material that neither emits nor reflects light, and to be able to pass right through ordinary material. (6)

While I must concede that there may very well be in (the) universe some "dark matter" which does not emit enough radiation to be detected, I cannot believe that there is 9 to 99 times as much of it as ordinary matter. Also, I do not believe that there is a "limited" amount of mass in (the) universe from which there is an awful lot missing. A recent article stated that astronomers currently estimate the total mass of "the universe" to consist of 125 billion galaxies. (I assume they mean by this, our region of (the) universe.) If we accept that (the) universe is infinite, then the amount of mass in (the) universe must also be infinite, since the average density (recently estimated at 6 hydrogen atoms per cubic meter in our region) is finite. And since I do not agree with the interpretation of the Hubble observations as evidence of an infinitely expanding universe, I do not believe that astrophysicists are on the right track to be looking for a fantastic amount of hidden ordinary matter to balance their questionable equations.

An even more mysterious form of super-dense matter, called Strange Matter, has been hypothesized, and some physicists believe it also may exist in substantial quantities throughout (the) universe and even in miniscule amounts, on Earth. Strange matter would occur beyond atomic number 109 where ordinary matter runs out, and consist not of protons and neutrons, but a soup of freely interacting quarks surrounded by a cloud of electrons, and would be extraordinarily heavy. A lump of strange matter the size of a BB would weigh more than 5 million tons. Charles Alcock and Edward Farhi suggest that this material could form in the core of neutron stars (remnants of supernovae) from the extreme pressure and temperature at which ordinary atoms could no longer sustain their structure — where electrons and protons fuse into neutrons — and then some of which would be further converted into free-moving up and down quarks. With the addition of enough up and down quarks, a seed of strange matter would be created at the core of the star. The seed would continue to grow by gobbling up all of the free neutrons until the star is essentially a strange star, enveloped by only a thin crust of ordinary matter. (Strange matter can't swallow up ordinary matter because the electric repulsion of their positively charged cores keeps them apart.) Strange matter has not been observed in space. Strange stars and neutron stars may appear identical, but a strange star would spin much faster due to its greater density. For a few hours in January 1989, astronomers caught a glimpse of a stellar remnant of a 1987 supernova which they thought might be a strange star. The star appeared to be spinning at about one-third the speed of light. It was reported that physicists also believe that they might be able to create strange matter in a particle accelerator, and such an experiment was to be undertaken at Brookhaven Laboratory. (7)

Some physicists think that the mass property of attractive gravitational force is the consequence of a mysterious discrete particle they call the Graviton. The graviton is viewed as a force-carrying particle (the conveyor of gravity) similar to the photon (the conveyor of light). A high density of gravitons flowing between bodies causes a high gravitational force. Our universe (called a brane) is considered to have a weak gravitational force relative to the other three natural forces of ordinary matter because few gravitons get through spacetime from another brane where they are concentrated. Dark matter (the 90% missing mass), some cosmologists suggest, is just ordinary matter concentrated on another brane, which cannot shine light through to our universe but whose gravitational force nonetheless is felt throughout our universe. (8)

The idea that antigravity particles (or material) might exist, or that such material might form complete planets or galaxies, is certainly not new. Such is the stuff of science fiction writers and children's imagination (e.g., Superman's magic Kryptonite, enabling him to at least neutralize gravity and leap over the tallest building or fly like a bird.) A cartoon originally appearing in Punch magazine in 1954, shows in the background, the entangled wreckage resulting from an accidental mid-air collision between a fighter type airplane and a flying saucer type spacecraft. In the foreground, the fighter pilot is shown parachuting down toward Earth, while the extra-terrestrial alien pilot is shown, inverted, parachuting upward, back toward outer space. (9)

In 1987, Paul Boynton, a physicist-astronomer at the University of Washington, published a study about "the most sensitive experiment conducted to date" to determine whether a Fifth Force exists — one which would slightly counteract gravity. The study reported that the observed motion of a bimetal ring consisting of equal semicircular masses of aluminum and beryllium hung beside a 400-ft granite cliff showed evidence of the fifth force. The aluminum half twisted toward the cliff while the beryllium half twisted away. The fifth force is believed to be related to the chemical composition, or makeup, of an atom rather than its mass. The strength of the antigravitational effect is related to a measure of composition called isospin, which is the number of neutrons minus the number of protons, and is so weak that it would be felt only over a range up to about 1000 yards. (10)

Note that the mysterious Fifth Force described above is an addition to the currently understood four natural forces of ordinary matter (attractive gravity, electromagnetism, strong and weak). It is not to be confused with another mysterious antigravitational force called Quintessence, or Dark Energy, named after Aristotle's fifth element, beyond earth, air, fire and water, "the etheric substance of heavenly spheres". In 1967 Yaakov Zel'dovich calculated that energy believed to be contained in the tiny elementary particles that occupy "the vacuum" (i.e., "empty space") of "our universe" amounted to 10^{120} times the total energy of the remaining "whole universe". Why this didn't blow everything apart — expanding faster than the speed of light was guessed to be prevented by "some sort of dynamic system which cancels out most of the energy of the vacuum, rendering it close to, but not actually zero". (11) Lawrence Krauss suggested about 1990 the name "quintessence" for the dark matter (missing mass) needed to hold galaxies and "our universe" together, but somehow it did not stick. The name now is applied to describe a variable energy constituent of space that represents a variable cosmological constant. Dark Energy is proposed as being a distinctive form of energy that unlike any other form of energy, which may be <u>converted</u> by a suitable transformational process into an equivalent amount of mass according to Einstein's equation, $E = Mc^2$, apparently has the property of equivalent mass which can exert gravitational influence. Ostriker and Steinhardt use the term to refer to a dynamical quantum field of gravitationally repulsive energy permeating space. Quintessence, unlike Einstein's vacuum energy that is considered to be completely inert and to maintain the same density for all time (i.e., time-invariable value of cosmological constant), "interacts with matter and evolves with time, so it might naturally adjust itself to reach the observed value today." (12) Quintessence as an energy field having negative pressure and repulsive gravity believed to spring from the physics of extra dimensions, is pushed out of galaxies to fill the intervening space. Ostriker and Steinhardt propose that the balance of forces in "our" expanding universe are met with a combination of ordinary and mysterious matters, consisting of only about 0.5% ordinary visible matter, about 3.5% ordinary but non-luminous matter, about 26% exotic dark matter, about 0.005% radiation, and the remaining predominant 70% dark energy (quintessence). They suggest that "if the acceleration is caused by quintessence...the universe might accelerate forever, or the quintessence could decay into new forms of matter and radiation, repopulating the universe." (12) Attempts to calculate the amount of dark energy residing in empty space (spread throughout "the universe") result in numbers 10^{60} times higher than astronomers have measured as needed to account for the expansion rate of "the universe" — enough to blow "the universe" apart before atoms and galaxies could form. Dr. Michael Turner says, "I think we are so confused that we should keep an open mind to tinkering with gravity." (13)

I believe the problem inherent in the idea of Dark Energy as an antigravitational constituent of space is to explain how it might act in a limited, localized fashion — to "be there" just when needed to exert control in a region of (the) universe where and when a big bang has occurred.

The term Negative Mass has appeared in several pieces of literature, mostly in passing conjecture about a form of matter that has never been found, but that some physicists believe should exist in (the) universe. (14) Although not a physicist (my field is thermodynamics), I too believe very strongly after many years of consideration that negative mass should exist in (the) universe. And furthermore, I believe it to be the most likely candidate for the mysterious matter that performs the basic function of providing gravitational counterbalance to ordinary matter (which for symmetry will be called positive mass), enabling (the) universe to be timelessly stable. While antimatter is considered to be the electrical mirror image of ordinary matter (all properties the same except electrical charge — opposite charges attract; like charges repel), negative mass is considered to be the gravitational mirror image of positive mass (all properties the same except gravitational force effect relative to "our positive mass" — opposite masses repel; like masses attract). Note that the designation of negative mass does not imply an opposite effect from the imposition of a force on basic inertial mass — pushing harder on negative mass would not cause it to slow down. And negative mass being electrically compatible with positive mass would not result in annihilation on contact. However, like contact between magnets, prolonged forced contact between positive mass and negative mass may result in the stronger one transforming the gravitational identity of the weaker one. And according to my hypothesis, negative mass is not a rare form of matter throughout (the) universe. It is only rare in regions of (the) universe during the time that positive mass prevails within those regions. On average, fifty percent of the mass in (the) universe must be positive mass and fifty percent negative mass. It is my fondest hope that someday an astrophysicist will discover some evidence of this material's existence.

QUOTATIONS

1. ...[A]ll modern attempts to include gravity with the other forces of nature in a consistent, unified quantum theory predict the existence of new gravitational-strength forces that, among other things, will violate...the equivalence principle of gravitation and the quantum-mechanical symmetry between matter and antimatter.

    Terry Goldman et al (2)

2. Perhaps the most exotic phenomenon between (and within) galaxies is the seething ocean of matter and antimatter pairs popping in and out of existence. This

peculiar prediction of quantum mechanics...can act as an antigravity pressure in the universe... .

Neil de Grasse Tyson (15)

3. We must regard it rather as an accident that Earth (and presumably the whole solar system), contain a preponderance of negative electrons and positive protons. It is quite possible that some of the stars is the other way about.　　　　　　　Paul Dirac, 1933 (16)

4. If antimatter galaxies were to encounter the small amount of matter outside a nearby matter galaxy, annihilation reactions would produce gamma rays of specific energy. The nonobservance of such gamma rays makes it unlikely that antimatter galaxies exist.

Robert Ehrlick (17)

5. Taken at face value, our observations appear to require that expansion is actually accelerating with time. A universe composed only of normal matter cannot grow in this fashion, because gravity is always attractive. Yet according to Einstein's theory, the expansion can speed up if an exotic form of energy fills empty space everywhere. This strange "vacuum energy" is embodied in Einstein's equations as a so-called cosmological constant. Unlike ordinary forms of mass and energy, the vacuum energy adds gravity that is repulsive and can drive the universe apart at increasing speeds... . Once we admit this extraordinary possibility we can explain our observations perfectly, even assuming the flat geometry beloved of theorists. Evidence for a strange form of energy imparting a repulsive gravitational force is the most interesting result we could have hoped for, yet it is so astonishing that we and others remain suitably skeptical.　　　　　　　Craig J. Hogan et al (18)

6. The need for dark matter began to insinuate itself into cosmology before World War II when...Jan Oort and...Fritz Zwicky noticed that galaxies behaved as

though they were far more massive than they appeared. ...Our own Milky Way [is] spinning faster than the laws of physics predict. Either we must demote our laws of gravity...or we must invent something that is holding galaxies together — unseen matter... . We can know it only by its secondary effects, phenomena that make no sense in the current theoretical framework unless we come up with more gravity, more mass. ...[F]urther measurements have suggested that... the ratio of the unseen matter to visible matter is ten to one.

George Johnson (19)

7. Scientists look to this energy [missing mass] to balance the books on cosmic mass, preserve their preferred model of the Big Bang theory of how the universe began, and predict how it will end. ...The trouble is, no one knows the identity and nature of the missing energy. Scientists, with a shrug of shoulders, speak of "something strange" or "funny energy". ...Dr. Michael S. Turner said "We want the missing energy to be there today and gone yesterday, in order to avoid interference with the growth of structure in the universe."

John Noble Wilford (20)

8. Dr. David Bennet [reported detecting]...the gravitational signatures of objects in the halo surrounding the Milky Way...[that] can be identified only by their gravitational force [and] are most likely to be white dwarfs [or]... extremely small black holes. ...They could constitute 50 percent of the mass in the Milky Way's halo...[and] 50 percent of dark matter in the universe... . ...Though some could be "something else" not yet explained. If they were normal glowing stars, they would be visible by ordinary means. John Noble Wilford (21)

9. [Jim Thomas says about the elusive "fifth force"]... "What we're really talking about is the possible modification of gravity, which is the fourth force."

John Langone (22)

10. Dr. Neta A. Bahcall...noted growing evidence that is "forcing us to consider the possibility that some cosmic dark energy exists that opposes the self-attraction of matter...".   John Noble Wilford (23)

11. Astronomers are increasingly confident that they have detected the first strong evidence that the universe is permeated by a repulsive force, the opposite of gravity. ...Dr. Michael S. Turner said... "Actually, the cosmological constant is the least interesting explanation, and that's pretty interesting in itself." ...Some weird form of energy may be influencing most of the universe.
                                        John Noble Wilford (24)

12. ...Some, as yet unknown, force operates in extragalactic space to keep the galaxies apart. ...No reason has been found why inert masses should not do what electric charges do, sometimes attract and sometimes repel others. ...Why does gravitation seem to contradict the Principle of Stabilization according to which every process in a self-contained system tends to reduce its cause?   Reginald O. Kapp (25)

13. ...[S]pace scientists, astronomers and physicists have... [no] explanation of a mysterious force that seems to be pulling spacecraft in the direction of the sun. ...[T]here is a slight possibility that some hitherto unknown phenomenon might be at work: "new physics", as physicists call such mysteries.Malcolm W. Browne (26)

14. ...[T]he universe may contain what is called "vacuum energy", energy that is present even in apparently empty space. By Einstein's famous equation, $E=mc^2$, this vacuum energy has mass. This means it has a gravitational effect on the expansion of the universe. ... Matter causes the expansion to slow down and can eventually stop and reverse it. On the other hand, vacuum energy causes the expansion to accelerate ... the opposite of that of matter.   Stephen Hawking (27)

15. Some physicists think gravity with a minus sign could exist someplace in the universe. Or that somewhere else out there could be negative mass that would have the effect of canceling out the gravity of positive mass like us.   John Boslough (14)

REFERENCES

1. Browne, Malcolm W. (1996, Jan5), Physicists Succeed in Creating Atoms Out of Antimatter, New York Times, p A9

2. Goldman, Terry, et al (1988, Mar), Gravity and Antimatter, Scientific American, p 48

3. Broad, William J. (1987, Jan 20), Search for Antimatter is Showing Results, New York Times, p C1

4. Lerner, Eric (1991), *The Big Bang Never Happened*, Times Book Div. of Random House, p 131

5. Wilford, John Noble (1994, Nov 1), Big Bang's Defenders Weigh Fudge Factor, A Blunder of Einstein's, As Fix for New Crisis, New York Times, p C1

6. Johnson, George (2000, Mar 5), Afloat in Cosmic Hall of Mirrors, New York Times, p 1wk

7. Vogel, Shawna (1989, Nov), Strange Matter, Discover, p 63

8. Johnson, George (2000, Apr 4), Physicists Finally Find a Way to Test Superstring Theory, New York Times, p F1

9. — (1955), *A Century of Punch Cartoons*, Simon & Schuster, p 114

10. (AP, Los Angeles) (1987, Sep 26), New Evidence Found for Fifth Force in Universe, Bennington (VT) Banner, p 9

11. Gorst, Martin (2001), *Measuring Eternity*, Broadway Books, New York, p 284

12. Ostriker, Jeremiah P. & Steinhardt, Paul J. (2001, Jan), The Quintessential Universe, Scientific American, p 46

13. Overbye, Dennis (2003, Nov 11), What is Gravity, Really?, New York Times, p F5

14. Boslough, John (1989, May), Searching for the Secrets of Gravity, National Geographic, p 563

15. Tyson, Neil de Grasse (1999, Jun), Between the Galaxies, Natural History, p 34

16. Broad, William J. (1987, Jan 20), Search for Antimatter is Showing Results, New York Times, p C1

17. Ehrlick, Robert (1994), *The Cosmological Milkshake*, Rutgers University Press, New Brunswick NJ, p 173

18. Hogan, Craig J., et al (1999, Jan), Surveying Space-Time with Supernovae, Scientific American, p 46

19. Johnson, George (1996), *Fire in the Mind*, Vintage Books, New York, pp 75 & 313

20. Wilford, John Noble (1998, May 5), Cosmologists Ponder "Missing Energy" of the Universe, New York Times, p F1

21. Wilford, John Noble (1996, Jan 17), Found: Most of Missing Matter Lost Around Edges of Universe, New York Times, p A1

22. Langone, John (1988, Aug 15), Was Sir Isaac All Wet?, Time, p 67

23. Wilford, John Noble (1999, May 26), Hubble Telescope Yields Data for Recalculating Age of Universe, New York Times, p A26

24. Wilford, John Noble (1998, Mar 3), Wary Astronomers Ponder An Accelerating Universe, New York Times, p F1

25. Kapp, Reginald O. (1960), *Towards a Unified Cosmology*, Scientific Book Guild, London, pp 162 & 169

26. Browne, Malcolm W. (1998, Sep 17), After Major Study, Mysterious Force in Space Remains Mystery, New York Times, p A21

27. Hawking, Stephen (2001), *The Universe in a Nutshell*, Bantam Books, New York, p 96

11

NEGATIVE MASS IN A DUAL-GRAVITY UNIVERSE

For those of us that opt for the steady-state existence of (the) universe, in opposition to the "Big Bang creation" hypothesis, there is a great question which must be addressed. Namely, How does (the) universe work to stave off an apparently ultimate collapse from the relentless effect of gravity?

In 1686 Sir Isaac Newton formulated the relationship between the gravitational force and the mass property of discrete bodies, which has come to be known as the Law of Gravity:

1) $F = GM_1M_2/d^2$ where:
   F = mutually attractive force
   M_1 = mass of one body in an interactive pair
   M_2 = mass of the other body in an interactive pair
   d = distance between the two bodies
   G = a constant of proportionality, called the Universal Gravitational Constant, measured by Cavendish (1731-1810) = 6.66×10^8 c.g.s.

You may note first of all that the force is persistent. I.e., it is independent of time. With all other factors remaining the same with respect to time, the force remains unrelenting and unchanged. That dependable force of gravity which among other things holds us to the earth's surface and the earth in its orbit around the sun, is certainly a stabilizing factor in our

129

personal existence. But its action on a celestial scale in other respects must be regarded as a destabilizing influence. During the natural course of celestial activity over time, such as the reduction of distance between two bodies which may be caused directly by the attractive force, or the increase in mass of a body caused by accession of smaller bodies, the effect is one of destabilizing positive-feedback, tending always toward the formation of bigger and bigger bodies:

1. Whenever two bodies in space react to the gravitational attraction force that exists between them by moving closer together, the force increases exponentially as the distance between them decreases, thereby accelerating the closure of the gap between them.

2. A large body in a field of smaller bodies may start to "swallow up" the closest smaller bodies, thereby adding their mass to its own. And with the increased mass, exert a larger gravitationally attractive force on other smaller bodies farther away, which would accelerate their motion toward the larger body.

The end-point of that unstable action, if there were nothing else happening to stop it, is one of the supposed "end of the universe" scenarios with everything being brought back together again (after the "Big Bang" explosion) into one big ball — or maybe a very tiny ball, collapsed into a singularity under its own massive weight.

However, if we believe that (the) universe has already been around forever, and we can see that it has not yet collapsed under its own weight, we find no reason to think that it will in the future. Therefore, I conclude that there must be something to stop it from doing so.

The purpose of this book is to encourage further thought in this field through the presentation of a speculative idea which I think is new. My hypothesis is simply that, if (the) universe is not going to collapse from the gravitational attraction effect that we so easily understand by observation, then there must be a self-regulating mechanism which involves a counteracting gravitational effect capable of working anywhere and everywhere throughout (the) universe — one of opposite nature (repulsion), which we have yet to clearly understand.

While the idea that an antigravitational material could (or should) exist in other regions of (the) universe (or even sparingly in our own Milky Way galaxy) is not new (1), what I offer as new is a consideration of the role that such material, called negative mass in the true Newtonian sense, might play in providing a gravitationally stable universe. The Law of Gravity provides no insight into a source of that phenomenon which appears to be an attribute of all matter, but it at least suggests that if the force acting between two ordinary (positive mass) bodies is one of attraction (+F), then the force acting between a positive mass and a negative mass would be one of repulsion (-F). And, again, the force acting between two negative masses would be one of attraction (+F). The following attempt to describe the operational role of negative mass, a material based only upon logic, imagination, and belief, grew out of conversations starting many, many years ago with a close friend, Jack Case, whose gentle nudging provided the encouragement for me to write all of this.

In preparing for this mental exercise, I contemplated some features of magnetic force, searching for similarities or analogies to gravitational force. First, I noted that the mutually attractive magnetic force acting between two magnetic bodies is polarized and directional. This effect is generally explained and illustrated as if the two bodies were made up of discrete miniature replications of themselves with proper polar alignment.

Fig. 1 Attractive Magnetic Force

Magnetic force can be created and controlled in a magnetizable material by application of an energetic field (e.g., an electric current in a coil, or exposure to an existing strong magnet). The field must have directional orientation as well as magnitude (a vector quantity) to produce the resulting magnetic vectorial force. Secondly, I noted that two magnets in juxtaposition will exert an influence on each other. Two magnets joined as shown in Fig. 1 will not only be attracted at their maximum force potential, but will mutually act to reinforce the maintenance of each other's strength. (I've kept a bar across the horseshoe magnet at my workbench, when it is not in use, for over 50 years, and it still seems as strong as when it was new.) However, if the magnets are forced together at similar poles, a stronger magnet can overcome and eventually reverse (transform) the polarity of a weaker magnet. The mutually attractive force of gravity acting between bodies in our known world of ordinary matter - positive masses - looks like this:

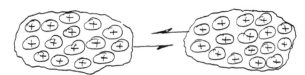

Fig. 2 Attractive Gravitational Force

While the resultant force acting between discrete homogeneous bodies is directed along the line passing through their centers (a vector force), their respective gravitational fields are uniform in all directions. (I weigh the same on my feet as I do standing on my head.) The gravitational effect is an isotropic property (one which is measured as uniform in all directions).

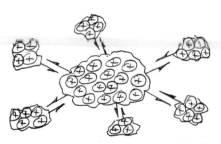

Fig. 3 Isotropic Gravitational Field

I believe that gravity is also a "created" property. And if so, it would logically be created by an isotropically energetic field — such as a region of high pressure and temperature. In this regard, I am pleased to learn that atom smashers have been used to create such an environment, momentarily, capable of producing antiparticles along with all of the more familiar particles. (2) Perhaps these antiparticles may someday be found to be related to antigravity particles. What truly creates the gravitational effect, or a proposed antigravitational effect, is a puzzle to me. But I would like to describe some ideas about how, once created, the gravitational identity of a body might be reversed.

The difference in the structure of material in a magnetizable bar between its magnetized and unmagnetized states is not evident even on microscopic examination. But at some point within the mass, at the atomic or subatomic level, there must be some difference between the material at the "(+) end" and the "(-) end" of the smallest element of the mass that can still be identified as a "magnet". If so, purely for illustration, I will label"(+) material = a" and "(-) material = b"

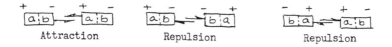

Attraction   Repulsion   Repulsion

Fig. 4 Magnetic Effect

Magnetic effect probably results from special orientation of particles within the atomic structure. From an experimental observation, we determine that "opposite" magnetic ends are attracted to each other and "like" ends are repelled.

In my concept of a gravitationally stable universe, there must be two different kinds of masses which have different gravitational effects on each other. Like magnets, they would attract or repel. Except we must concede as we can see in our own world of positive masses (A+/A+), that "likes" attract. Therefore, "opposites" (A+/B-) repel.

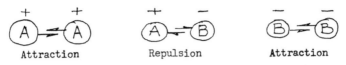

Fig. 5 Gravitational Effect between Positive and Negative Masses

I would like to interject at this point that my concept of negative mass is that, in its own world, it is a very real material. Robert Wright says "...[T]he structure of our world depends on pattern, not on the pattern's substrate; a carbon atom according to Fredkin, is a certain configuration of bits, not a kind of bits." (3) Carbon and Iron in a "negative" world should have the same physical properties as in our "positive" world — and maybe the strawberries even taste the same.

But if negative mass is a real material, why hasn't it ever been found in our world as a result of the latest big bang or of more recent meteoric action? The best answer that I can give is that the existence of such a body on our planetary surface would be only momentary because if it were not restrained it would be repelled back into space. Less obvious perhaps, but vital to my concept is the belief that if a body of negative mass were actually trapped within the material that eventually became our earth's shell from the time than an actual big bang created our solar system, the hunk of negative mass material would be slowly transformed into positive mass.

It is further tempting to think that negative mass might be involved in the seemingly inexplicable catapulting of asteroids from beyond Mars, which are currently explained as the origin of meteorites resulting from unexpected gravitational instabilities. (4) It is also tempting to suggest that such things as the 1908 Siberian Explosion site or other craters in which no trace of meteoric material can be found, might have resulted from a collision between the earth and a piece of negative mass material that had more than enough inertial impetus to continue its path away from a big bang explosion in a neighboring "negative" galaxy into our Milky Way, even against a repelling gravitational force.

Just as a single mode of attractive gravity (positive or negative) works relentlessly to shape all discrete celestial bodies into nearly perfect spheres, a self-regulating dual-gravity system capable of maintaining a stable universe must work relentlessly toward spatial uniformity — a nearly perfect grid work of alternately repulsive bodies. As an idealized model, a balanced universe of discrete positive and negative bodies that is neither expanding nor contracting from the dual-gravitational effect would appear as in fig. 6 — except three-dimensionally, and extending infinitely in all directions, of course.

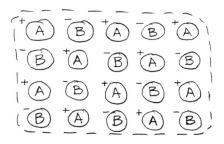

Fig. 6 Ideal Universe, Gravitationally Balanced

Naturally, such a perfect arrangement would be unstable. It would tend to change with time due to unavoidable positional perturbations as the positive masses became increasingly attracted to each other and the negative masses became increasingly attracted to each other and the positives and negatives more repulsive of each other. And if that's all there was to it, we would end up, not with one big ball, but two big balls as shown in fig. 7.

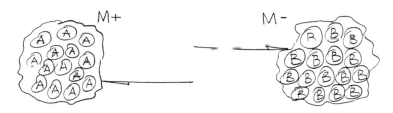

Fig. 7 Ideal Universe, Gone Awry

135

So the next thing that my stable universe model must include is an allowance for bodily accretion due to gravitational attraction (making big ones out of little ones), but only up to a point (something must keep black holes from growing indefinitely). Once that point of maximum growth for a discrete body is reached, a mechanism would be needed not just to stop further growth, but to regenerate the broad range of particle/body sizes that make up the cosmic dust, asteroids, moons, planets and stars that we know exists now throughout (the) universe (making little ones out of big ones) and redistribute them back into space, and at the same time reverse the gravitational identity of the remnant body to restore gravitational balance (equal attraction and repulsion forces) to the neighborhood. But what would it take to have such a system capable of oscillations and reversible processes?

Before getting into that, let me describe in more detail the imagined "broad-brush" difference between elemental units of positive and negative mass. How could two pieces of elemental mass, otherwise composed of the same material, differ only in their gravitational identity? I know that there are a lot of different things that can be "done" to atomic material to produce various elements or isotopes — adding or removing electrons, protons, neutrons, pions, muons, etc., etc. — but after all that is done, what else could be done to any element to produce a different gravitational identity? Of course, I don't know that anything could be done. But if there could under the right conditions, it might be something fairly simple.

At subatomic levels, gravitational forces seem to play an insignificant role in particle activity. Perhaps at that level the gravitational effect has not yet been "created". Can it be that gravitational effect is a result of the way that particles are combined, and not just in the numbers of each kind? It is generally understood that the planetary model of atomic structure of an element (electrons orbiting around a nucleus of neutrons and protons) is not a very accurate description of

136

the real thing. However, I will represent the structure of an element of positive mass as shown in fig. 8. Then an element of negative mass could be as shown in fig. 9.

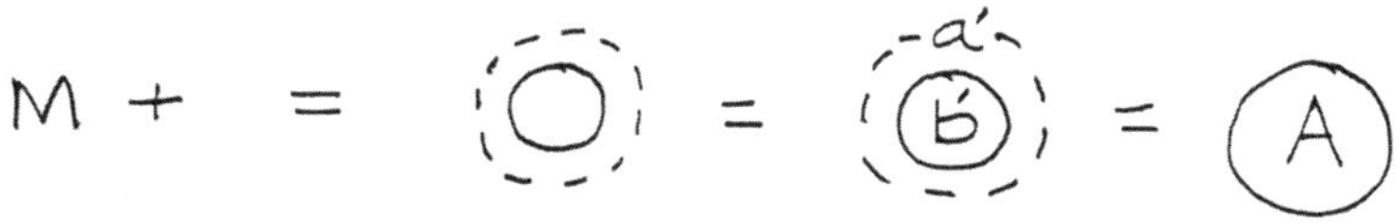

Fig. 8 Schematic Diagram of Positive Mass

Fig. 9 Schematic Diagram of Negative Mass

With these in mind, let's consider M+ and M- as being comprised of basically the same material (a' and b'), but orientated in different ways. I'll call the transformation from one to the other the "popcorn effect". To describe how this might happen, we must lean on the discipline of quantum mechanics, which was developed when earlier methods of observing and interpreting atomic and subatomic particle activity began to break down. Gary Zukav says: "What we have been calling matter (particles) is constantly being created, annihilated and created again. This happens as particles interact and it also happens, literally out of nowhere. Where there was 'nothing' there suddenly is 'something', and then the something is gone again, often changing into something else before vanishing. In particle physics there is no distinction between empty, as in 'empty' space, and not-empty, or between something and not-something. The world of particle physics is a world of sparkling energy forever dancing with itself in the form of its particles as they twinkle in and out of existence, collide, transmute and disappear again." (5) With buzz-words of "random" and "probability", quantum mechanics has come to describe how anything that can happen, may happen sometime under certain conditions at some level of probability.

The super-microscopic world in my model could be pictured like a miniature, never-ending demolition derby, with particles acting like cars colliding — losing a fender in one crash, picking up a fender in another. The pace of the race and the frequency of collisions would depend on the state of the environment. In a moderate environment, the race would proceed unendingly, with a "Chevy" (an M+ particle) still looking pretty much like a Chevy — especially if it were only racing with a bunch of other Chevys. But in an extreme environment, where the severity and frequency of impact is significantly higher, it may be after a Chevy has been knocked about badly, that just ever so often it looks rather odd — more like a "Ford" (an M- particle). That's what I mean by the "popcorn effect". The extreme conditions that promote more aggressive levels of particle activity are well known to be higher pressure and temperature. Fantastically high levels of pressure and temperature are required to achieve the process of atomic fusion. In nature, the place to look for these conditions would be in the core of a huge celestial body.

Fred Hoyle and Willy Fowler described in a 1960 paper, summarized by Simon Mitton, the implosion of a dying star, as follows:

> "At the very end of the life of a massive star, the material in the nuclear core has reacted to make the elements of the iron peak, which are the ultimate products of energy-releasing fusion reactions. All fusion reactions of nuclei heavier that these require an *input* of energy, so the iron core can draw on no further source of energy to replace the vast amounts being emitted at the surface of the star. Inexorably, the gravitational force is going to overwhelm the dying nuclear furnace. As the core cools, its internal pressure falls, and its ability to remain stable against the inward crush of gravity diminishes. The tip moment comes very suddenly: Hoyle and Fowler showed that once the nuclear fusion peg is kicked away, the central regions of a massive dying star implode catastrophically in just one second! ... The

energy generated by this infall is immediately converted to heat, which manifests itself by greatly increasing the speeds of the protons (hydrogen) and helium nuclei. Suddenly, there is a nuclear explosion that processes as much of these light elements in one second as the Sun consumes in a billion years. Explosion of the outer layers of the star follows the implosion of the core, which in turn releases a vast amount of gravitational energy in its collapse. This is the standard picture of supernovae explosions today, first described by Fowler and Hoyle." (6)

The largest bodies currently imaginable are the invisible black holes whose gravitational fields are believed to be so strong that even light cannot escape them. The pressure and temperature that we can create with such difficulty and for so brief a time for a nuclear fusion must be mere "child's play" compared to conditions in the core of a dying black hole, and perhaps there is more going on than we currently imagine. What really happens at the core of a huge mass?

We know that the hydraulic pressure acting on a body under the surface of our earth increases in proportion to depth. This is a gravitational phenomenon due to the weight of the material above the body. Temperature is also known to increase as a function of this "adiabatic compression", and from the nuclear reactions going on in the core. But what is the nature of hydraulic pressure/gravitational force acting on a small body at the core of a large body, since "above" is no longer definable? Newton's law indicates that the force of attraction between two bodies approaches infinity as the distance between them approaches zero. But that does not apply for one body inside another. How then does this force relate to stress or pressure acting on a smaller body when the other body is all around it? I picture the core region of a huge mass as one in which the bodies/particles would float as in a zero gravity free space — in an "ish" of supercritical fluid — maybe even a region actively hostile to the very structure of matter. Thus any body

139

or particles trapped on/in a growing larger body on the scale of a black hole may be put through a pressure and temperature cycle of some tremendous range, ending up in an environment ideally suited to the creation of atomic mutations. Could this be a process for seeding and incubating a core of "different material"?

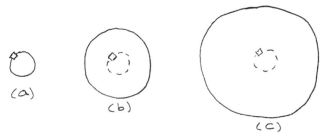

Fig. 10 Body growing by accretion

If the transformation of a "Chevy" (M+) particle into a "Ford" (M-) particle were possible, but only at a very low level of probability in some marginally energetic environment, then the chances of any Ford meeting another Ford before it got turned back into a Chevy again would also be very low. But if the possibility existed, and the environment were energetic enough to turn Chevys into Fords with a high enough probability so that a small group of Fords could get together through their mutual attraction for each other and repulsion by the Chevys, then a "beachhead" could be formed.

Imagine a pregnant balloon — an otherwise 100% M+ black hole — trying to self-abort an unwanted fetus — a core of M- material (Fig. 11). Certainly the core would feel engulfed by gravitational rejection, and it would in turn begin to exert a reactive force on its surroundings, thereby establishing a clear-cut "battleground" or "no-man's land" of interplay around the core (Fig 12).

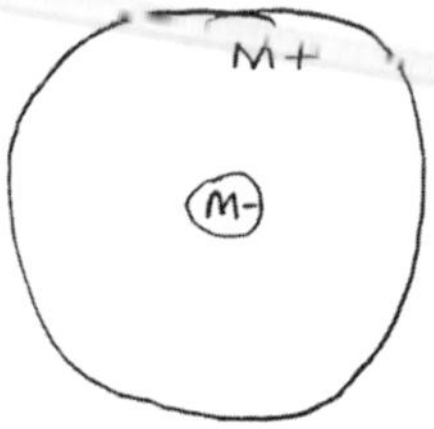

Fig. 11 "pregnant" celestial body

M+

(M-)

140

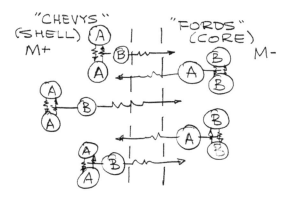

Fig. 12 No-Man's Land

Next, let's look at the concept of this "no-man's land" in the demolition derby model, remembering the high probability of Chevys being turned into Fords, Fords being turned back into Chevys, and each side capturing about-equal gains (Fig. 12). We might question if the core really has a statistical chance to grow, since the probability of a Ford being transferred out of Chevyland should be about the same as a Chevy being transferred out of Fordland. However the nature of a definable core and shell is such that the surface area on the shell side of the "no-man's land" interface will always be greater than the surface area on the core side. Thus, there would always be a higher number of Chevys available to be turned into Fords than vice versa. So the core should continue to grow.

As its core grows, the "pregnant balloon" should begin to develop some distinct "labor pains". A massive body in its last throes collapses inwardly from the accumulated excess weight, increasing the pressure and temperature of its core, and spinning faster and faster from conservation of angular momentum. At some point, the repulsive force between the M- core and the M+ shell could become high enough to act in combination with the increasing centrifugal force to fracture the shell and send its fragments flying into space — leaving behind a remnant body of pure M- material.

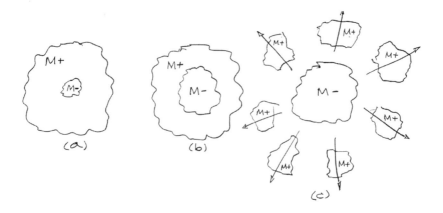

Fig. 13 Is this a Supernova?

Stephen Hawking describes the typical life of a star as including an initial period of "burning up" available fuel in the form of atomic reaction, followed by a period of contraction which would convert helium into heavier elements like carbon or oxygen. "What happens next is not completely clear, but it seems likely that the central regions of the star would collapse to a very dense state, such as a neutron star or black hole. The outer regions of the star may sometimes get blown off in a tremendous explosion called a supernova, which would outshine all other stars in its galaxy." (7)

A recent article in Scientific American describes a Quasar as "... a 'powerhouse' — probably a rotating black hole many millions of times more massive than the sun — at the center of a starry galaxy." (8) Another article in the same magazine describes a Pulsar as "... one gigantic atomic nucleus, held together by gravity rather than the nuclear strong force; a thimbleful scooped from its interior and transported to the earth would weigh 100 million tons. An international team of astronomers says it has glimpsed this fantastic object — a rapidly spinning neutron star, or pulsar — nestled deep within the expanding envelope of supernova 1987A. Theorists have predicted that supernovas leave behind a neutron star, but none of their models envisioned a remnant like this one." (9)

The fact that black holes are observed as numerous and discrete bodies throughout (the) universe tells me that they do not grow forever. And they do not appear to be rushing wildly towards one another, even with their unimaginably overpowering gravitational strength, which prevailing wisdom tells us should make them mutually attractive. But maybe they aren't all attracted to each other. How else, than by having a duality of positive and negative masses throughout (the) universe can we explain black holes swallowing up all local "debris" — but only occasionally merging with each other, and in some cases surrounded by supernovae (blasting material outwardly)? Why didn't they all converge long ago? Why are we still here? To me, it is evident that there is a natural mechanism at work throughout (the) universe which involves a balancing of gravitational forces between such bodies, not evident on a local planetary/intragalactic scale but active on a celestial intergalactic scale. From Earth, a long-term observer of activity in the vicinity of the imagined black hole, (fig. 13), would have noticed some strange things happening. While it was once a huge mass sucking into itself all nearby smaller bodies of M+ and repelling all others of M-, (Fig. 14a), it would appear later to have lost its tremendous gravitational force, perhaps start glowing again, and the vicinity around it become calmer before the approaching storm — the big bang (Fig. 14b).

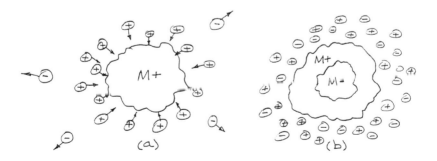

Fig. 14 The Calm Before the Storm

After the big bang, the remaining core of M- would quickly repulse any smaller M+ bodies lingering in the vicinity and begin gathering up the available M- bodies to grow eventually into an M- black hole. To complete the cycle, that black hole would ultimately develop a core of M+ that would eventually blow the M- shell off, and the remaining M+ core would then sweep up the smaller M+ bodies around it to become an M+ black hole again. However, if this sort of thing happens at all, it would be more realistic for it to occur at different times in varying degrees of intensity over a wider range of conditions than portrayed above — not always starting out as dramatically as in a black hole and ending in a supernova. Also it should be noted that during the transition period that the body was composed of two different materials, M+ and M-, the net gravitational force extended to other distant bodies must be diminished by the interaction of the gravitational forces of the composite materials. If all of this is possible — that positive and negative masses may be combined for a time within a single body — then Newton's formulation of the Law of Gravity, 1), must be a special case of a more general formulation:

2) $\quad F = G(A_1 - B_1)(A_2 - B_2)/d^2$ where:

$\quad\quad F =$ gravitational force (+ attractive) (- repulsive)

$\quad\quad\quad\quad\quad A_1 =$ amount of M+ in body 1

$\quad\quad\quad\quad\quad B_1 =$ amount of M- in body 1

$\quad\quad\quad\quad\quad A_2 =$ amount of M+ in body 2

$\quad\quad\quad\quad\quad B_2 =$ amount of M- in body 2

$\quad\quad\quad\quad\quad$ (G & d as defined in eq. 1)

Four special cases would be:

2.1) $\quad F = G(A_1)(A_2)/d^2 = +$ max. attractive force between positive masses. (B_1 and $B_2 = 0$) (Newton's Law of Gravity)

2.2) $\quad F = G(A_1)(-B_2)/d^2 = -$ max. repulsive force between a positive mass and a negative mass. (A_2 and $B_1 = 0$)

2.3) $\quad F = G(-B_1)(-B_2)/d^2 = +$ max. attractive force between negative masses. (A1 and A2 = 0)

2.4) $F = G(0)(A_2-B_2)/d^2 = 0$ no gravitational force between two bodies due to one having combined equal amounts of positive and negative mass. (e.g., $A_1=B_1$)

In this manner, or something very close to it, I believe that (the) universe is balanced at any given time by a scattering of positive stars and galaxies and negative stars and galaxies. They all would look alike from a great distance and could only be classified according to their gravitational identity across the wide spectrum of strongly-positive to neutral to strongly-negative by observing the way that neighboring groups of bodies interact. Certainly, discrete bodies (the small stuff — cosmic dust to stars, including solar systems) should be one or the other, positive or negative. But the giant stars — the black hole or neutron star approaching supernova status — might very well be into the transition phase. Black holes are reported to be "gobbling up" matter like crazy while at the same time ejecting matter. And there are those stars seen to be drifting aimlessly in space far removed from any galaxy, which might indicate that they have a diminished or total lack of gravity. There are other stars seen to come "out of nowhere" that suddenly become the brightest light in the sky before exploding as a supernova, which may indicate a black hole that has suddenly lost its light-withholding gravitational force in the process.

Whole galaxies of stars should be predominately one class or the other (positive or negative masses), especially for smaller galaxies having no black hole or just a single black hole. However, it is likely that larger galaxies with several black holes (of the same gravitational identity) would begin to fragment when one of the black holes reached the critical point of growth and changed its gravitational identity before the other black holes got around to it. And even in a smaller galaxy with only a single black hole at its center there might be residual "other" mass bodies. The time involved between cyclic reversals of the gravitational identity of the central body may not be long enough for the galaxy to be completely cleansed of the "old" smaller bodies which surrounded an earlier central body (before its last big bang), by the time that the next one happens. E.g., there may still be some "negative mass" bodies being ejected from the Milky Way following its last big bang (15 billion years ago?) that transformed it into a "positive mass" galaxy.

All of this I speculate to happen as a result of the chance encounters of subatomic particles in a special extreme environment of bodies that we really know nothing about, starting a chain of events like the famous butterfly described in James Gleick's *Chaos* that flutters its wings on one side of our planet and sets off a wind disturbance which grows and travels to the other side resulting in a hurricane, that changes the structure of stars and galaxies, recharges energy fields, regenerates and redistributes the basic building material of matter, and keeps (the) universe gravitationally balanced for all time.

QUOTATIONS

1. [C]osmologists Paul Steinhardt of Princeton University and Niel Turok of University of Cambridge,... proposed that the big bang is not a one-of-a-kind event but part of a recurring cycle. ... Gabriele Veneziano of CERN...says, "Thanks partly to the work of Turok, Steinhardt and colleagues, our community is much more ready to accept that the big bang was the outcome of something rather than the cause of everything."

   George Musser (10)

2. The Hubble Space Telescope by chance photographed the exploding star, the most distant ever observed, in 1997. Scientists say subsequent detective work on the relative intensity of its light confirms one of Einstein's conjectures about the universe: that all space is bubbling with an invisible form of energy that creates a mutual repulsion between objects normally attracted to each

other by gravity. ... Cosmologists will have to cope with a universe that seems increasingly filled with mysterious stuff that scientists cannot see and do not fully understand. James Glanz (11)

3. Where did...mass come from? The mass comes entirely from the arrangement of the quarks and not at all from the quarks themselves. Dennis Overbye (12)

4. To understand oscillation, first know dualism. ...[An unidentified Japanese scientist said] "We say that Buddhist tradition makes it easier for us to grasp the dualities and transformations of particle physics. For example, the duality of particles and waves manifested in all the ingredients of matter has long been known, but although everyone accepts this duality intellectually, the idea may be hard to feel in your heart..." Malcolm W. Browne (13)

5. ...[A] galaxy [is]...perhaps the most natural unit of matter on the cosmic scale. J. J. Callahan (14)

6. ...[G]alaxies [appear] to develop from the bottom up, not from the top down. That is, they begin as smaller objects, ...and grow into larger galaxies and groups of galaxies. John Noble Wilford (15)

7. ...[A]stronomers said they were surprised that Hubble's spectrograph has detected interglactic clouds of hydrogen in regions relatively near the Milky Way. Dr. Stephen Haran [said]...they are not sure if the observed hydrogen clouds are halos around undetected galaxies, or isolated structures that failed to evolve into galaxies, or else remnants from immense explosions accompanying galaxy formation. John Noble Wilford (16)

8. Gravity alone, which scientists had thought was too weak to do the job, could have moved matter about to account for the universe's rich texture of clustered galaxies interspersed with gaping voids. John Noble Wilford (17)

9. ...[S]cientists have found evidence that supermassive black holes probably lurk at the core of nearly all galaxies and the mass of each one seems to be proportional to the mass of the host galaxy. ...Dr. Douglas Richstone [says]..."this is probably the key evidence that black hole's presence is intimately connected to the evolution of the galaxy." ...Dr. Mitchell Begelman and Sir Martin Rees...concluded "when we really understand black holes, we will understand...the universe itself."

John Noble Wilford (18)

10. Not all elliptical galaxies necessarily contain black holes. Even if a black hole once resided in a galaxy, it may no longer be there. Martin J. Rees (19)

11. These reports, together with somewhat more circumstantial evidence that our galaxy itself contains a central black hole, suggest these fantastically compact objects may play a crucial role in the evolution of many — and perhaps all — galaxies. After the black hole consumed most of the matter within its gravitational grasp, [Ramesh] Narayan says, the quasar might finally evolve into a less luminous, more normal-looking galaxy

John Horgan (20)

12. ...[C]areful optical measurements of soft X-ray novae in their quiescent state have given the most convincing evidence of the existence of stellar-mass black holes inside our galaxy. G. S. Bisnovatyi-Kogan (21)

13. Astronomers have detected a titanic explosion in the outer reaches of the cosmos — one so violent and bright that for about 40 seconds it apparently outshone all the rest of the universe. Except for the Big Bang that is generally believed to have created the universe, no other cosmic explosion of such magnitude has ever been discovered. Science has no models that could produce an explanation for so stupendous an outpouring of

energy. ...[Dr. Stanley E. Woosley said] "I'm a very troubled theorist. ... We're really struggling to find a theoretical basis for this." ...Dr. Alexei V. Filippenko ...suggested that the burst could have been caused by the violent merger of two black holes.

Malcolm W. Browne (22)

14. No principle of physics determines how massive a black hole can be.   Jean-Pierre Lasota (23)

15. ...Dr. [Stephen] Hawking...[says that black holes] leaked tiny particles of energy into space, slowly shrinking and heating up until the black holes finally expired in a fiery explosion... [which] he whimsically called thunderbolts.

Frederic Golden (24)

16. [Quantum] theory will cause black holes to radiate and lose mass. It seems that they will eventually disappear completely... .   Stephen W. Hawking (25)

17. ...[T]he pressure at the Earth's center has been estimated to be 3.6 million times the atmospheric pressure at the surface. Under such enormous pressure, solid iron would have to have a temperature in excess of 7000 K. Therefore, the core of the Earth must actually be hotter than the surface of the Sun.   Robert Ehrlick (26)

18. Astronomers have observed...nearby stars that appear to be orbited by planets more or less the size of Jupiter. Astrophysicists could only speculate how such large planets, ranging in mass from one-half to 10 times the mass of Jupiter, can orbit so close to a star and survive what must be destabilizing gravitational stresses. ... They might be some new type of object unlike anything in the solar system. "To find a new type of object would be an extremely exciting result," Dr. [William D.] Cochran said. "Right now we are still going to call them 'planets', even though they may really be something slightly different."   John Noble Wilford (27)

19. [Dr. Geoffery W. Marcy, Dr. Michel Mayor and Dr. William D. Cochran wrote] "This pile-up of planets near stars suggests that inward orbital migration occurred after formation," probably through gravitational interactions with other planets that, in reaction, were nudged outward into distant orbits where they are not yet detectable.　　　　John Noble Wilford (28)

20. Supernovas are important to galactic ecology: they cook light atoms into heavy ones and then spew them back into space; all the gold in the universe, among other elements, originated in this way.　Timothy Ferris (29)

21. ...[A]s [dying stars] heated up further... [they] would not release much more energy, so a crisis would occur... . ...[I]t seems likely that the central regions of the star would collapse to a very dense state such as a neutron star or a black hole. The outer regions of the star may sometimes get blown off in a tremendous explosion called a supernova... . Some of the heavier elements produced near the end of the star's life would be flung back into the gas in the galaxy, and would provide some of the raw material for the next generation of stars.
　　　　Stephen W. Hawking (30)

22. [Renata Kallosh says] that some extreme black holes can detonate, releasing three hundred times the energy of a nuclear fission bomb.　　　Timothy Ferris (31)

23. It has not been clear how exactly mass from the outer parts of a galaxy is transformed to the black hole at its core, although its powerful gravity obviously plays a dominant role. ... Though the findings show that all matter on the brink of a black hole is not doomed, scientists cannot explain what is going on. Somehow, soon after the ejection of hot gas from the inner disk, more matter fills the void and the process is repeated. Scientists suspect this is a common occurrence in the vicinity of small black holes. John Noble Wilford (32)

24. Astronomers think [that]...[b]ecause smaller galaxies [which may outnumber giant galaxies, like the Milky Way] have less of a gravitational grip on their contents, the dispersed gases from their first stars to explode may vanish into intergalactic space... . ...Dr. [John] Kormendy noted... that the smallest dwarf galaxies are not shreds torn from colliding large galaxies, as has been theorized, but "are real galaxies." John Noble Wilford (33)

25. ...[A] star, Wolf-Rayet 104, ...is spewing gases while at the same time rotating around an unseen stellar companion. Madhusree Mukerjee (34)

26. For the first time, astronomers have captured an image of a cloud still in the process of collapse. Gas and dust is falling into the young protostar, which is simultaneously emitting narrow jets of similar material. ... The apparent inside-out collapse has been a source of confusion for theorists and observers alike. More than a decade ago, astronomers began finding that nearly all newborn stars go through a phase in which they seem to be rejecting mass at the same time they were also presumably drawing mass from the cloud collapse. John Noble Wilford (35)

27. A new calculation suggests that dense star clusters, which are sprinkled throughout galaxies in Earth's cosmic neighborhood, act as assembly lines for tightly orbiting pairs of black holes and then spit them out of the cluster. There they eventually fall together and merge into one. ...[D]etectable mergers of black holes could be as much as a thousand times more common that the neutron stars... ...[The] black holes attract each other gravitationally and pair off, forming binaries. The binaries carom off other black holes and eventually pick up enough speed to escape the cluster. Out in the dark wasteland of space, the black holes spiral inward just as

neutron stars do, and they finally merge in a spectacular collision.   James Glanz (36)

28. Dr. Henry C. Ferguson...said today that this was the first time stars had been detected more than 300,000 light years ...from the nearest galaxy. ...[S]ome stars do not have a galaxy to call home. Somewhere along the way, they wandered off or were tossed out of the galaxy of their birth. Many more such homeless stars presumably exist in the region but are too dim to see.
                              John Wilford Noble (37)

29. Astrophysicists representing the Naval Research Laboratory, Northwestern University and the University of California at Berkeley...announced [the discovery of] what appears to be a monster fountain of antimatter erupting outward from the core of the Milky Way. ...[T]he cause and nature of the antimatter fountain were puzzling. It might be a continuous shaft of antimatter streaking northward from the galactic center, or it might be a cloud, separated from the main part of the galaxy.
                              Malcolm W. Browne (38)

30. Probabilistically speaking, it is mind-bogglingly more likely that everything we now see in the universe arose from a rare but every-so-often-expectable statistical aberration away from total disorder, rather than having slowly evolved from the even more unlikely, the incredibly more ordered, the astoundingly low-entropy starting point required by the big bang.
                              Brian Green (39)

31. The new conceptions of black holes eliminate the event horizon altogether The basic idea is that there does, in fact, exist a force that could halt the collapse of a star when all else fails. That force is gravity itself. In matter with certain properties, gravity switches from being an attractive force to a repulsive force.
                              GeorgeMusser (40)

REFERENCES

1. Boslough, John (1989, May), Searching for the Secrets of Gravity, National Geographic, p 563

2. Broad, William J. (1987, Jan 20), Search for Antimatter is Showing Results, New York Times, p C1

3. Wright, Robert (1988, Apr), Did the Universe Just Happen?, The Atlantic Monthly, p 29

4. Gleick, James (1987), *Chaos*, Penguin Books, New York, p 314

5. Zukav, Gary (1979), *The Dancing Wu Li Masters*, William Morrow and Co., Inc., p 212

6. Mitton, Simon (2005), *Conflict in the Cosmos*, Joseph Henry Press, p 242

7. Hawking, Stephen W. (1988), *A Brief History of Time*, Bantam Books, p 40

8. Horgan, John (1989, Apr), Points of View - Quasars and Radio Galaxies May be Two of a Kind, Scientific American, p 20

9. Horgan, John (1989, Apr), Points of View - It's a Pulsar, Scientific American, p 22D

10. Musser, George (2002, Mar), Been There, Done That, Scientific American, p 25

11. Glanz, James (2001, Apr 3), Photo Gives Weight To Einstein's Thesis Of Negative Gravity, New York Times, p A1

12. Overbye, Dennis (2001, Mar 20), In New Physics, No Quark is an Island, New York Times, p F1

13. Browne, Malcolm W. (1998, Jun 9), Room in the Universe for Ancient Belief and Modern Physics, New York Times, p F4

14. Callahan, J. J. (1988-89?, ?), The Curvature of Space in a Finite Universe, Scientific American, p 90

15. Wilford, John Noble (1998, Nov 24), A New Baby Picture of the Southern Sky, New York Times, p F5

16. Wilford, John Noble (1991, May 18), Gases May Reveal Planet Formation, New York Times, p 8

17. Wilford, John Noble (1990, Feb 1), A Computer Model Suggests Gravity Shaped the Universe, New York Times, p B8

18. Wilford, John Noble (1997, Jan 14), New Findings Suggest Massive Black Holes Lurk in the Hearts of Many Galaxies, New York Times, p C1

19. Rees, Martin J. (1990, Nov), Black Holes in Galactic Centers, Scientific American, p 56

20. Horgan, John (1987, Oct), Hearts of Darkness, Scientific American, p 30

21. Bisnovatyi-Kogan, G. S. (1998, Feb 27), At the Border of Eternity, Science, p 1321

22. Browne, Malcolm W. (1998, May 7), Astronomers Detect Immense Explosion 2nd Only to Big Bang, New York Times, p A1

23. Lasota, Jean-Pierre (1999, May), Unmasking Black Holes, Scientific American, p 40

24. Golden, Frederic (1992, Jun 9), Theory and Whimsy Take Physicists on Tour Through a Black Hole, New York Times, p C1

25. Hawking, Stephen W. et al (1996, Jul), The Nature of Space and Time, Scientific American, p 60

26. Ehrlick, Robert (1994), *The Cosmological Milkshake*, Rutgers University Press, New Brunswick, NJ, p 196

27. Wilford, John Noble (1999, Jan 10), In New Discoveries, A Planetary Mystery, New York Times, p 18

28. Wilford, John Noble (1999, Apr 20), In a Plethora of Planets, New Fields of Science, New York Times, p F1

29. Ferris, Timothy (1998, Aug 10), Seeing in the Dark, The New Yorker, p 55

30. Hawking, Stephen W. (1988), *A Brief History of Time*, Bantam Books, p 120

31. Ferris, Timothy (1995, May 15), Minds and Matter, The New Yorker, p 46

32. Wilford, John Noble (1998, Jan 13), Peak at Black Hole's Feast Reveals Awful Table Manners, New York Times, p F1

33. Wilford, John Noble (1999, Jan 7), Dwarf Galaxies May Hold More Dark Matter, New Studies Show, New York Times, p A19

34. Mukerjee, Madhusree (1999, Jun), Stellar Pinwheel, Scientific American, p 18

35. Wilford, John Noble (1995, Oct 10), In a Dust Cloud, Rare Glimpse of Star Aborning, New York Times, p C1

36. Glanz, James (1999, Oct 26), On a Galaxy's Assembly Line, Where the Black Holes Merge, New York Times, p F4

37. Wilford, John Noble (1997, Jan 15), Hubble Detects Stars That Belong to No Galaxy, New York Times, p A12

38. Browne, Malcolm W. (1997, Apr 29), Enormous Plume of Antimatter Alters View of the Milky Way, New York Times, p A1

39. Green, Brian (2004), *The Fabric of the Cosmos*, Alfred A. Knopf, New York, p 176

40. Musser, George (2003, Jul), Frozen Stars: Black Holes May Not Be Bottomless Pits After All, Scientific American, p 20

12

GRAVITATIONAL LENSES IN A DUAL-GRAVITY UNIVERSE

If we accept the concept of a black hole as a body of such huge mass that its gravitational force contravenes the transmission of light from what must surely be a very hot body, then we must accept the photon theory of light which involves its having a minuscule measurable mass and associated ability to react to a gravitational force. Personally, I remain skeptical of the claim that light rays may be withheld or deflected by gravitational force, simply because I have not had the opportunity to study whether other alternatives which seem obvious to me have been fully considered and found lacking. Certainly I understand the implications of stars seen to be orbiting around invisible partners and whole galaxies rotating around an invisible core. I firmly believe in the existence of black holes as gravitational monstrosities because they are the logical result of long-term gravitational attraction and accretion of bodies having common gravitational identity within a proximate space. But might it be possible that the gravitational monster has a surrounding atmosphere of "cosmic dust" that is so thick (possibly millions of miles deep) and so densely packed, as to simply create an opaque screen around the "black hole" body? If so, it still might be argued that given the amount of heat being generated within the body, that even if the screening atmosphere were packed as densely as a metal, like iron, then the screen itself would become incandescent and radiate light. But if the screening atmosphere were thick enough to be opaque and at the same time of a low enough density to act as an insulating blanket — keeping the heat in while maintaining its own outer region at a temperature below the incandescent point by radiation to the black void — the body would be invisible to us for a completely different reason.

And also imagine what would happen to such a body — possibly nothing more than a super-heated neutron star with an opaque shroud of "cosmic dust", rotating rapidly — when its "gravity switch" is turned off (i.e., a negative mass core has grown enough to weaken its gravitational hold onto the shroud). The shroud of "dust" would quickly dissipate, leaving the star exposed — naked — to shine brilliantly, just before the big bang.

I am also skeptical about the "gravitational lens" effect reported in many articles, whereby the light rays radiated from a distant source (galaxies or quasars) are believed to be deflected and focused by gravitational force upon passage around/ through a large body/galaxy, as predicted by Einstein in 1915. Since then, observations have been made of bright sources that appear to be deflected in space or seen as an "Einstein Ring". But while some articles claim good agreement between observed light deflections and the prediction of deflection of light passing near large bodies based on the theory of general relativity, Petrosian (1) estimates an error magnitude of 10X between an observation and prediction, with the difference being attributed to the effect of "dark matter". I do not doubt that the rays of light from a distance source may be deflected upon passing close to a large body or through a galaxy, but this could also be the result of optical refraction through the large body/galaxy's atmosphere of "cosmic dust", just as sunlight is refracted in our own atmosphere of "dirty" air (e.g., sunsets and rainbows).

However, if the theory that light responds to gravitational force is valid, then we must address the question of how this fits into the idea of (the) universe being composed of an equal amount of two different kinds of materials having differing gravitational identities. If half of the bodies/galaxies in (the) universe are predominately positive mass and the other half are predominately negative mass, then what would be the effect of light from a distant positive mass source passing through the gravitational field of a closer negative mass — a very probable occurrence? First of all, we would have to acknowledge that there must be two different kinds of light — "positive light" which would be radiated from positive mass bodies and "negative light" which would be radiated from negative mass bodies. And following the same gravitational rules for other masses, "positive light" would be deflected toward a large positive mass body, but would be deflected away from a large negative mass body, and vice versa for "negative light".

Thus, if the "gravitational lens" phenomenon is actually caused by gravitational force and if (the) universe is composed of both positive and negative mass bodies, then we should be able to observe a duality of gravitational lens effects. The simplest effect of creating an image of the light source shifted in space is illustrated in Fig. 1. Normally when we look at a "star" in the night sky we have no way of knowing whether the light rays from that source have been deflected one way or another by an intermediate large body. But if we could focus on a particularly bright and distant source — well outside of our own galaxy — and wait a long time (a few billion years??), we might see the apparent position shift around dramatically as the light rays reach us after spacetime travel — having passed at one time near a large positive mass body and at another time near a large negative mass body. Or would we only see the last shift caused by the large body nearest to us — a positive mass body in the Milky Way? Within our lifetime, of course, we can only observe time-variable displacements within our own solar system (like eclipses) or the Milky Way galaxy, which does not leave much chance for the last deflection of light rays to be caused by a negative mass (outside of the Milky Way) before reaching the earth.

The case of light rays forming an "Einstein Ring" after passing through a galaxy or around a large body and focusing on the earth is a little more interesting. This phenomenon would only appear when the light source and the intermediate body have the same gravitational identity as shown in Fig. 2a. No "Einstein Ring" image of the light source would appear focused on Earth if the gravitational identities of the source and the intermediate lensing body were different, as shown in Fig. 2b. I.e., we simply would not be aware of an alignment of a light source and a large body which fails to produce an otherwise expected lens effect.

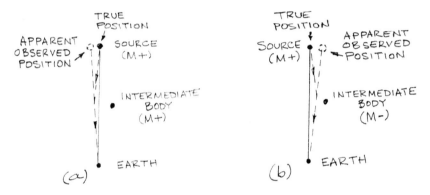

Fig. 1 Gravitational Displacement of Light

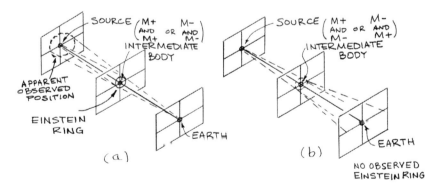

Fig. 2 "Einstein Ring" Effect

159

QUOTATIONS

1. Gravitational lenses are particularly well suited to...the study of dark matter in the universe. ...[I]nvestigators have steadily accumulated evidence of large-scale gravitational fields that are far stronger than can be explained by the observed (that is, luminous) stars and interstellar material. Most astronomers interpret this to mean that 90 to 99 percent of the universe's total mass is made up of some undetected component — the postulated dark matter. Edwin Turner (2)

REFERENCES

1. Horgan, John (1988, ?), Cosmic Mirage, Scientific American, p 22

2. Turner, Edwin L. (1988, Jul), Gravitational Lenses, Scientific American, p 54

3. Wilford, John Noble (1990, Jan 19), Gravity is Used in a First "Look" at Distant Matter, New York Times, p A16

4. Browne, Malcolm W. (1995, Apr 18), Dark Matter Search Hints at New Shape for Galaxy, New York Times, p C1

5. Browne, Malcolm W. (1998, Mar 31), "Einstein Ring" Caused by Space Warping is Found, New York Times, p F3

6. Wilford, John Noble (1998, Dec 29), The Universe as Telescope, New York Times, p F1

7. Tyson, Neil de Grasse (1999, Jun), Between Galaxies, Natural History, p 34

13

WHAT NEXT?

I would like to think that my writing will help to form a more believable narrative about the nature of "our universe" (i.e., (the) universe) — like striking a match to light a candle which may dispel the darkness. But whether the candle stays lit or gets blown out will depend upon others whom I have been able to persuade to keep it alive. I am realistic enough to anticipate that most professional astronomers and astrophysicists (the die-hard Big Bangers — if they read this at all) will be unimpressed, if not appalled, with my ideas and continue on more serious quests, like hunting for the graviton (the "particle" that conveys the force of gravity from one body to another — similar to the photon, conveyor of light), the Higgs field (the "ocean" of resistance to accelerating motion that imparts the properties of mass and inertia to individual bodies), and Loop Quantum Gravity (a theory about the discrete structure of spacetime on the smallest of scales and the related structure of matter bridging the gap from Einstein's theory of relativity to quantum mechanics).

But I would humbly ask that at least some of the scientific leaders reconsider some of the most puzzling celestial observations by looking at them again through a "negative mass lens" — not only to resolve a question about whether or not "negative mass could cause that kind of behavior", but if affirmed, to establish some sort of definitive measure to "prove" the existence of "negative mass" in other regions of (the) universe.

Otherwise, I continue to look for reports of celestial events that may show the "footprint" (i.e., possibility) of "negative mass". I am pleased to find more recently, books and articles arguing that the Big Bang was nothing more than a big bang, and that time and space really did exist before "the Big Bang" — even if it is described as dragging "the universe" through a knothole.

Following are some examples of recent reports that I find intriguing:

The Ring Around the Milky Way Galaxy

At the winter conference of the American Astronomical Society, held in Seattle during January '03, astronomers learned about a newly discovered ring of hundreds of millions of stars reaching all the way around the Milky Way galaxy in a nearly circular orbit. (1) Some of these stars had been seen faintly in the 1990s from Mount Stromlo Observatory in Australia, but were mostly hidden by the stars, dust and gas in the galactic plane. It took the extended effort of two teams of astronomers based in New Mexico and the Canary Islands to map the stars and recognize their extent. The ring, at about 120,000 light-years diameter, is oriented basically in the galactic plane, though it is about 10 times as thick as the rest of the galaxy, and extends well above and below the galactic plane — somewhat like a balloon tire on a racing bike's wheel. Within the Milky Way, stars are more numerous and spaced closer together near the center and thin out towards the edge. But now we are aware of this distinctive ring farther out, and what are we to make of it? What are they doing out there, almost "by themselves", and how did they get there? The two possibilities are that, 1) they came from outside, or 2) they came from inside the Milky Way galaxy.

1) The conventional view is that the ring was formed by collision of our galaxy with smaller or dwarf galaxies and the resulting mix-up. But is this the logical pattern of collision between galaxies having the same

gravitational identity (both "positive mass")? There are much earlier reports of smaller galaxies just "passing through" the Milky Way. If this ring is the result of a collision with a smaller galaxy, I would suggest as a definite possibility that the smaller galaxy was a "negative mass" galaxy whose stars are repelled by, rather than attracted to, the composite body of the Milky Way galaxy. The ring-shaped formation may be retained by the mutual attraction between those stars in the ring creating a circumferential bond with just enough force to balance against the radial force of gravitational rejection from the center of the galaxy — similar to the elastic force that keeps a rubber tire on its rim at high rotational speeds.

2) Dr. Annette Ferguson, a member of the Canary Island observation, offered a possible alternative explanation. "The stars might have come from inside the galaxy's disc..., and gravitational interactions could have disturbed their orbits, causing them to migrate outward and into the ring." Fine! But how exactly do you get 500 million stars to suddenly migrate out of a galaxy — only to stop and hover around its outer edge? (And at this point, I should emphasize that however those stars got there, it is much more probable that it was the result of a cataclysmic event — e.g., a big collision or a big internal blow up — rather than a slow accumulation of stars arriving one-at-a-time over a long time period, each to stake out a claim on "empty space" and await the future arrival of others of its own kind.) This looks like for a job for Supermass (a.k.a. "negative mass")! To me, this puzzling structure seems very much like another example of how the dual-gravity scenario might be manifested. The latest big bang within "our universe" (estimated at 10 billion years ago when our galaxy was formed — which I believe transformed it

from a "negative mass" galaxy to a currently "positive mass" galaxy), would have sent lots of "negative mass" stars and shell fragments flying out, leaving a "positive mass" core to attract other smaller "positive mass" bodies into the region. And over a long period of time it would build up the current Milky Way galaxy — now estimated to contain 400 billion stars. As the "negative mass" material was blown outward, forced by the blast, light pressure, rotation, and gravitational rejection, most of it probably disappeared into the void of space. But at some distance, the repelling gravitational force could have relaxed enough before some of the bodies lost their communal chain of attraction to hold together circumferentially. As the new Milky Way grew larger with time, attracting more and more "positive mass" matter available in the region, the ring would have been pushed farther out like an elastic waistband.

A Star Leaving the Galaxy

A star traveling through the Milky Way at 1.5 M miles per hour was seen recently by astronomers from the Harvard-Smithsonian Center for Astrophysics. Since that speed is twice the escape velocity for the Milky Way's gravitational field, the star is presumably "flying the coop" — already 6 times farther from the galactic center than our solar system. Conventional thinking is that it is the remnant of a binary star pair that passed close enough to the black hole at the center of our galaxy, for the pair to be torn apart by the enormous tidal forces. One star was sucked into the hole and its momentum was transferred to the other, which sailed by with a near miss at increased speed. (2) My alternative explanation is that the star, designated SDSS J090745.0+24507, might be a "negative mass" star that was shot into our galaxy from a big bang not far away, and finding itself in the "wrong neighborhood" was evicted — like a tennis ball bouncing off a barn door.

How do Supernovas Happen?

Dennis Overbye recently described in a New York Times article (3) the current state of thinking among leading scientists about supernova explosions — "…one of the most complicated things in the big wide world." The basic quandary seems to be a lack of understanding and agreement about just how a supernova explodes and what triggers such an event that may be seen somewhere in the sky once a second or so, blossoming briefly to the brilliance of a billion suns. A long-held notion is that in a binary star system, one of the stars accumulates matter from the other, and when a critical mass is reached in the accumulating star, it explodes. "But details of that explosion, which happens invisibly in a second or so, are still a mystery. … The problem is that there are two ways for a star to burn: like a flame, which is called deflagration, and as an explosion, a detonation in which the burning propagates as a shock wave. And neither type of burning, by itself, can easily explain what astronomers have seen in supernova explosions." One model adopted by theorists is that the star burns in the flame mode initially, slowly expanding, and then detonates when the star reaches a critical density and chemical composition. But "none of these 'delayed detonation' models explains why or when the star would detonate." Dr. Donald Lamb says: "It turns out that you need walls to have an explosion", but a star has no walls. Dr. J. Craig Wheeler said that finding a natural trigger for the detonation is the "silver chalice" of his profession. An idea that the flame could start to burn slightly off-center in the star, and then sweep around it and slam into itself with enough force to set off the detonation looks good on a computer screen, but is considered an unlikely event in the case of a real star. And I would add, if all that isn't complicated enough already, imagine the idea of adding the creation of a "negative mass" core in the star to grow and set off the detonation.

For you, the reader, I would hope that my writing provides some measure of logic and conviction toward the idea that (the) universe should be considered as infinite, and therefore such terms as "the creation of", or "the age of", or "the size of", or "the shape of", or "the missing mass of", or "the multiplicity of", or "the end of" (the) universe, simply lose any meaning. Demand some clarity in the usage of terms — especially the word *universe* — by using in your own writing the term *(the) universe* to designate an infinite universe (as opposed to "the universe" which is a region of (the) universe. And write letters to editors demanding the same. And then realize that an infinite universe has no other choice but to be stable and timeless. Then imagine the most logical way that (the) universe might operate to maintain overall stability in spite of localized chaos. And if something better than the idea of the existence and role of "negative mass" comes along, promote it.

Become a Neo-Steady Stater. The label of Steady Stater for those choosing to not believe the premises of the Big Bang hypothesis, has fallen into disrepute since such stalwarts as Albert Einstein and Sir Fred Hoyle had to admit that the evidence of expansion going on in "the universe" was strong enough to show that "it" was not static. And likewise, the native of a South Seas island would have to admit that his "universe" was not static, after being swept clean by a once-in-a-thousand-years tsunami. But we're no longer talking about the older, limited "universe(s)". This is about (the) "new" (infinite) universe — the one that was there before the "old" one was "created" in the mind of man.

QUOTATIONS

1. The idea of an antigravitational force pervading the cosmos does sound like science fiction, but theorists have long known that certain energy fields would exert negative pressure that would in turn, according to Einstein's equations, produce negative gravity. ... (Dr. Robert Caldwell added) "Our calculations show, however, that galaxies reside in a bubble of old-fashioned Einstein gravity, whereas gravity has changed outside and between galaxies." Dennis Overbye (4)

The Nature of (the) Universe

2. The simplest conclusion is that some as yet undiscovered physical law causes the cosmological constant to vanish. ... Two interpretations are viable: either the universe is open, or it is flat by some additional form of energy that is not associated with ordinary matter. ... I believe the evidence points in favor of the latter, [but] either scenario will require a dramatic new understanding of physics.

Lawrence M. Krauss (5)

3. The world, you might argue, does not need yet another subatomic particle. But even particle physics has not been about particles for a long time, physicists say. Rather it is about relationships between particles, the symmetries that nature seems to respect, in short, about the beauty that physical laws seem to embody. ... In the 30 years I have been following this stuff, it has never been wilder. But the real best seller here is wonder. It was popular science books that reminded me finally why I had been interested in science. There's a whole universe out there, and nobody knows how or why.

Dennis Overbye (6)

REFERENCES

1. Wilford, John Noble (2003, Jan 14), In Galaxies Near and Far, New Views of Universe Emerge, New York Times, p F3

2. Overbye, Dennis (2005, Feb 22), Going, Going, Gone! A Star Leaves the Galaxy, New York Times, p F1

3. Overbye, Dennis (2004, Nov 9), Life-or-Death Question: How Supernovas Happen, New York Times, p F1

4. Overbye, Dennis (2004, Feb 17), From Space, A New View Of Doomsday, New York Times, p F1

5. Krauss, Lawrence M. (1999, Jan), Cosmological Antigravity, Scientific American, p 53

6. Overbye, Dennis (2004, Jul 27), After Triumph and Disillusionment, Wonder Re-enters the Story, New York Times, p F3